Upholstery
Techniques Illustrated
2nd Edition

Upholstery
Techniques Illustrated
2nd Edition

W. Lloyd Gheen

TAB Books
Division of McGraw-Hill, Inc.

New York   San Francisco   Washington, D.C.   Auckland   Bogotá
Caracas   Lisbon   London   Madrid   Mexico City   Milan
Montreal   New Delhi   San Juan   Singapore
Sydney   Tokyo   Toronto

Ultracel is a trademark of Arco Chemical Company

SECOND EDITION
FIRST PRINTING

© 1994 by **TAB Books**.
TAB Books is a division of McGraw-Hill, Inc.

Library of Congress Cataloging-in-Publication Data

Gheen, W. Lloyd.
    Upholstery techniques illustrated / by W. Lloyd Gheen. — 2nd ed.
       p.   cm.
    Includes index.
    ISBN 0236224 ISBN 0236232
    (pbk.)
    1. Upholstery.   I.   Title.
    TT198.G48 1994   93-44951
    684. 1'2 — dc20   CIP

Acquisitions Editor: Kimberly Tabor
Managing Editor: Lori Flaherty
Editor: Robert Burdette
Production: Katherine G. Brown
               Rose McFarland, Desktop Operator
               Ruth Gunnett, Computer Artist
               Tina Sourbier, Typesetting
               Lorie L. White, Proofreading
Design team: Jaclyn J. Boone, Designer
               Brian Allison, Associate Designer
Cover Photo: Bender & Bender, Waldo, Ohio   TAB1
Cover design: Theresa Twigg   0236232

Contents

REUPHOLSTERY TECHNIQUES

List of tables

Introduction

Upholstery Techniques Illustrated, 2nd Edition, has been revised with two major purposes in mind. The first is to provide a sequential and rather complete how-to primer for novices. The second purpose is to serve as a reference for those taking upholstery classes and for the do-it-yourselfer. This second edition contains information on cleaning upholstery fabrics and detailed information on selecting and using foam rubbers.

As with any technical book, there are some new terms as well as some not-so-new terms for which basic assumptions could lead to confusion. Therefore, I suggest you consult the glossary as you read so you know what is really meant. As one example, what is the difference between *upholstery* and *reupholstery*? Now, you might be inclined to assume that "re-" anything is the redoing of something that has been done before, and "upholstery" is doing whatever it is the first time. Check it out in the glossary so you'll know what I am really saying.

Extensive photographs and illustrations are included to provide a step-by-sequence to get the job done. Tricks-of-the-trade are included to help make the work successful. Most tools, materials, and procedures that have been discontinued in modern upholstery practices have been omitted. Technical information on the newer materials and tools has been expanded to help you make informed decisions rather than rely exclusively on my bias.

Upholstery is so diverse that one reading of this book or doing one piece of furniture will not make you an expert. Each piece of furniture brings its own new challenges and problems to be solved. This is an ever-changing and exciting trade. New and different tools, materials, and procedures are constant realities, and this book cannot be the "last word." I welcome constructive criticism and suggestions for additions and improvements.

Part I

SUPPORT INFORMATION

Part 1 contains two fundamental types of information. The first is general—information that does not pertain to any particular style or piece of furniture but is fundamental for a basic understanding of upholstering furniture. This includes selection and cleaning of fabrics, tools, a basic classification of furniture styles, alternatives to finishing off a piece of furniture, and sewing machine care and maintenance.

The second is categorically specific—aspects that might apply to almost any furniture but not necessarily required for any particular piece. Included is the application of buttons, making channels, tufting, and working with vinyls. If you know what information is wanted, turn to the chapter addressing that topic. For instance, suppose some covered buttons have popped out of a couch and all you want to do is to install some new covered buttons. Turn directly to chapter 7, Buttons and channeling.

1
Overview

Throughout the book, two generic terms appear: upholstering and reupholstering. *Upholstering* usually refers to the application of all new materials to a new frame, which could include any or all of the following: foundation, base, padding, stuffing, and cover. Used more loosely, it means *reupholstering*, application of new cover material to used upholstered furniture. This may or may not include new foundation, base, padding, or stuffing. Whenever *upholstery* is used in this book, it means reupholstery unless specifically noted otherwise.

Upholstering new furniture or reupholstering old furniture remains more an art than a technology. Although there are technical aspects that will lead to a finished unit that is pleasing and long lasting, very few upholsterers do a job exactly the same way. For example, variations in the way stuffing and padding materials are applied give different "feels" to the furniture and result in different appearances of the finished product. Some like the tight, rather squared look, while others may prefer the more subtle, soft, "puffy" look. Both effects can be achieved using the same materials. Small changes in how and to what extent the cover is fitted and pulled down can transform a piece from the tight to the softer look, and vice versa.

Based on my experience, there is no such thing as one way of upholstering. However, to say that there are no procedures better than others would have about the same validity as saying that because a stork often stands on one leg rather than both, he has no need of the other.

Many beginning upholsterers have encountered some frustration trying to understand exactly how to do a particular part of a job when they have been told one way of doing it and another person steps up and corrects their procedure, saying, "That's not the way to do that. Here, let me show you how." The question naturally arises, which is the right way? The answer? Hold on to your hats. This is profound! Common sense and good judgment are often the most valuable guides. Strange thing about those two characters, though—common sense is not so common, and good judgment is often debatable. Just remember—whatever the boss says is the best way. What seems to work best and is easiest for *you* is probably the soundest guide, coupled with a background of the "basics," of course. This text is designed to provide those basics and some of the tricks of the trade.

For the novice, upholstering possesses many challenges and problems not previously confronted. "Where does this go? How should this be done? What now?" All these are common queries. So accept a bit of counsel at this point, and don't get all hung up in finite details.

Upholstering is a problem-solving activity. Do some thinking. Try to envision what would happen if that task was done as you imagine it. If you think it over and

Don't get all wrapped up in *finite* details . . . .

it seems logical and nothing would be seriously wrong if done your way, try it. But don't fail to think.

Upholstery is fun! Some who have tried it, however, have said, "I'll never, ever do this again!" Or "Now I can see why they charge so much to reupholster a chair." But there are also those who after taking one or more classes, have started their own businesses, hired into existing shops, or continued doing every piece of upholstered furniture they could in their homes or for family members. These latter types found upholstering fun.

One vivid and true example of the joy possible in upholstery can be seen from this illustration: A slight, energetic, pleasant, elderly lady beyond her sixties' not only did her own work, lifting and moving her chair around, making and correcting mistakes; she also made a lengthy trip to buy an antique love seat, stripped it down, refinished the wood, and reupholstered it with diamond tufting and deep piped channels using a striking deep red, heavily napped velvet (not an easy undertaking for the novice)—and all because she enjoyed it so much.

Taking a class in upholstery or just doing a piece of furniture on your own is guaranteed to accomplish four things:

1. Tax your patience.
2. Increase your tolerance to frustration (if the job is completed properly).
3. Reveal some muscles that have long since been forgotten (overlooked?).

4. Develop a confidence (that may have been concealed or lacking before) that you can tackle another piece and do it well if you desire.

Think it through before doing it.

Note: You cannot become a competent upholsterer with one experience. Every piece of furniture seems to have its own personality that requires a slightly different approach and possesses a few unique problems not previously experienced. Several experienced upholsterers (each having over 27 years in commercial upholstery) I esteem as top-quality journeymen commented that they still learn with each new piece that comes into their shop. One of these men has worked with fabric costing more than $200 per yard (it had gold thread in it); has done production and custom work; has redesigned and designed furniture from the frame up; has worked draperies, sporting, and personal equipment that can be sewn or repaired on a commercial straight-stitch sewing machine; and has designed, with patents pending, athletic and gymnastic gloves and wrist supports in use at universities, high schools, hospitals, and therapy centers in the United States—and all this with no commercial advertising.

What does all this mean? Simply that the first, second, or third time you upholster on your own using this book will not give you all the skills and all the answers to problems encountered on different types of furniture. Upholstering (reupholstering) is a problem-solving activity, not a science.

The first time you strip a unit to the bare frame to make frame repairs or replace badly worn or soiled materials, just looking at that skeleton of what used to be a piece of furniture is enough to instill panic and then despair. Take courage! All is not lost! If the stripping process is pursued as outlined in the second part of this book, a pleasing restoration can take place.

Proceed now into the fascinating, challenging and rewarding world of upholstering. Remember—even when your finger is dripping blood because you got caught on a staple stub you forgot to get out during the stripping process, it can be fun. Even if your patience is taxed, your muscles grumble at you because they haven't been used in quite this way for a long time, your hair thins (from pulling), and your blood is shed on those "gotchas"—even after all this—it can be fun. *Buen aventura!*

2
Upholstery tools & their uses

The tools discussed in this chapter are organized into three basic categories: (1) essentials for the do-it-yourselfer; (2) the minimum for the small or part-time shop of, the serious hobbyist; and (3) tools for the professional shop. An attempt has been made to list the tools for each category in a priority order, the most essential first. This basic ordering might be altered by circumstances. One person might find that every tool is essential; another, that some tools not mentioned in this book should have been. The categories are intended only as a guide.

Although tools are often used for many purposes, the proper use of the appropriate tools is presented in this and subsequent chapters. The safe way is to use tools for the purposes for which they were made. Take the time to find and use the right tool, and the work will be a pleasure with a high chance for success. But, taking shortcuts just because something else is handier was the source of the modern upholsterer's proverb: "He who takes shortcut with tools finds longest and most rough way."

FOR THE DO-IT-YOURSELFER

Upholsterer's shears After trying to cut upholstery fabrics, no one would consider doing it with a pair of common household scissors. Heavy-duty *upholsterer's shears* are necessary. If a lot of cutting is to be done at one time, shears will be much more "friendly." They don't create blisters nearly as fast or muscle cramps nearly as acute as common scissors.

Screwdrivers Disassembly and reassembly of some furniture pieces would be virtually impossible without the right screwdriver. Several specialty screwdriver types that have come into use, but for most assembly, the *straight-slot* and *Phillips* remain most popular. A medium size would probably be most practical for the one-time upholsterer. However, the only way to preserve driver blades and screw heads is to use the right size driver for each screw size. If the driver fits snugly, it is the right size. If it doesn't quite enter the slot or cross, it is too large. If it only goes halfway across the screw slot or wobbles without turning the screw, it is too small.

Use the right tool for the right job.

Staple removers & ripping tools Only a glutton for punishment would think of stripping a piece of furniture without a *staple remover* for stapled materials, a *ripping tool* or *claw tool* for materials fastened with tacks (FIG. 2-1). If a claw tool or ripping tool is to be used, some hammering device is necessary. Only a karate expert would use a bare hand to drive these tools. A straight-slot screwdriver could be used to remove staples and tacks, but that is a torturous way to go.

The *ripping tool* (FIG. 2-1A) is probably best for starting the removal of tack and tacking strips. With its flat bottom, wedge-shaped blade, and offset handle, it gives a smooth entry and good lift to get tacks or staples clear of the frame wood. For removing tacks, use the *claw tool* (FIG. 2-1B). This tool also has an offset handle for efficient wedging action; the V slot in the blade is intended to straddle the tack. Lifting the tack this way tends to pull it out straight. The ripping tool can give lift to only one side of a tack, which will often bend it over rather than lifting it out.

Staple removers are the prime tools for removing staples. The *Bostitch (blade) puller* (FIG. 2-1C) is an excellent tool for thick fabrics and soft woods. The hardened steel blade (replaceable) is slipped between the legs of the staple, then rocked backward, lifting the staple out. To use this tool for staples in hard woods (oak) is frustrating for most upholsterers. They would prefer the *Osborne (square-shanked) puller*, which has two sets of sharpened, V-shaped prongs at its tip. The bottom prong on one side can be pushed under the most stubborn staples and with a twist of the wrist, pull out at least one leg of the staple. Pliers or dikes can then be used to remove the other leg. Probably the most popular of the staple removers is the *Berry picker* (FIG. 2-1E), named after its inventor. Although this is the most expensive of the five staple removers; it seems to be the fastest and most versatile.

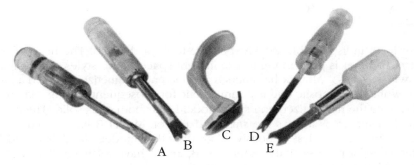

2-1 Stripping tools: (A) ripping tool; (B) claw tool; (C) Bostitch (blade) puller; (D) Osborne puller; (E) Berry picker.

Measuring tape Some flexible measuring device is necessary. Two popular types are the *steel tape* and the *cloth tape*. Flexibility is essential, since many measurements must be made around curved surfaces and corners. One word of caution: Don't try to communicate to another a measurement using the cloth tape unless you remember that the actual measurement begins at the point of the arrow marked *start* and not at the darkened end of the tape. The added length is provided so the operator can have something to hold on to when measuring to corners, edges, and the like.

Pliers A great aid in removing staple remnants is a pair of pliers. Handy pliers are the *slip-joint, diagonal cutters* "dikes," and *needle-nose*, shown in FIG. 2-2A, B, and C, respectively. Each has a specialty use in addition to performing the functions of pulling, cutting, and bending. *Slip-joint* pliers are especially useful for general grasping and bending, such as edge wires and other light spring wires. I have used a large, 10-inch pair for removing thousands of staples. Sometimes, the rocking motion that can be used with these pliers is easier on the wrist than the Berry picker. Dikes are the stripper's helper. They can grasp very close to the frame surface, cut staples easily when they can't be pulled out, and are flat enough to be used quite successfully as a "gotcha" (staple-stub) finder. The *needle-nose* is able to make small-diameter bends in light wires and get into some places too tight for fingers and other types of pliers to reach.

Tack hammers Although most modern upholstering is done with staples and staple guns, all of the *tacking* can be accomplished with upholstery tacks. The *tack*

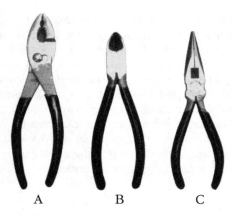

2-2 Popular pliers: (A) slip-joint; (B) diagonal cutters ("dikes"); (C) needle-nose.

hammer is an indispensable yet inexpensive tool for that job. The head of a standard claw hammer is just too big to use in tight spaces. Four styles of tack hammer are shown in FIG. 2-3. All tack hammers have one end magnetized to hold a tack for starting, while the opposite end is nonmagnetic for subsequent driving. Hammers A, B, and C have the magnetized end split. Model D has both ends solid. The magnetic end on all hammers is always the smaller diameter. Model B has less curve to the head, less taper, and is less expensive than the other models. Models A and D are preferred of most professionals. Models A, B, and C have relatively flat sides, used by many upholsterers to set and level *tack strip*.

Adhesives Furniture frames occasionally need repairs or modifications. Good-quality *wood adhesives* are frequently used. The most popular adhesives for wood-working are the white polyvinyl glues and the yellowish aliphatic. The aliphatics

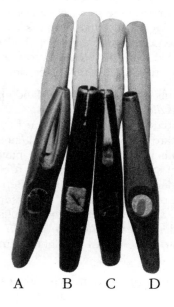

2-3 Tack hammers: (A) split-end magnet (greatest curvature of both heads); (B) split-end magnet (least curvature of both heads); (C) split-end magnet (straight-end for magnet, curved head); (D) solid-end magnet (moderate curvature, both heads).

generally give a higher-quality bond. However, even the polyvinyl glues, when properly applied, will give a bond stronger than the wood itself.

Other adhesives necessary for the upholsterer are the *foam adhesives*. For the small-volume operator, these are most popularly aerosol. For high-volume use, (i.e., furniture manufacturing), they can be purchased in bulk and applied with spray guns. Two varieties of foam adhesives are helpful: (1) an adhesive to attach foams to substrates such as wood, metal, cork, or other fabrics (a general-purpose contact adhesive) and (2) a less expensive variety such as the foam and fabric adhesive. When foam, fabric, or carpet is to be applied to areas where high heat (100–140° F) is a factor, like the inside of vans, trucks, etc., something with higher tack (stickiness), like super trim adhesive, is preferred.

FOR THE SMALL SHOP

Although upholstering can be performed successfully with the minimal assortment of tools described above, doing more than one unit is made much more enjoyable and profitable with additional tools.

Staple gun   An indispensable tool for today's serious upholsterer is the *staple gun*. Figure 2-4 shows various types of guns. For the average homeowner who doesn't have an air compressor handy, the *manual staple gun* (FIG. 2-4A) can be used. If the manual gun is desired, get the heavy-duty models that have two tension settings. Lighter, inexpensive models do not have the power to drive staples into furniture woods. The *electric staple gun* (FIG. 2-4B) is quiet and more bulky than its more popular counterpart, the pneumatic (FIG. 2-4D). The electric gun is very handy for site repairs where hauling around an air compressor would be impractical. Heavier, wider staples are generally used in the electric gun than in pneumatics, which sometimes is a disadvantage. The *pneumatic gun,* the workhorse of the industry, is used extensively in vir-

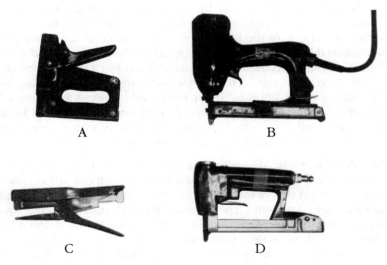

A   B

C   D

2-4   Staple guns: (A) manual, spring loaded; (B) electric; (C) scissor, also manual; (D) pneumatic, the workhorse.

tually all professional shops and educational programs. For most applications, 60 psi is sufficient pressure. The *scissors stapler* (FIG. 2-4C) is ideal for all temporary stapling: tucks and pleats in cover panels (preparatory to sewing or fastening to the unit), the edges of dacron (when wrapping cushion foam), and other light-duty work.

Clamps   The *C-clamp* (FIG. 2-5A) and the *spring clamp* (FIG. 2-5B) are used in applications requiring pressure in small areas, such as holding parts steady for drilling, holding parts together while glue is setting, or clipping fabric out of the way temporarily while working beneath it. The C-clamp is capable of exerting near crushing pressure should it be necessary; the spring clamp exerts a rather mild pressure.

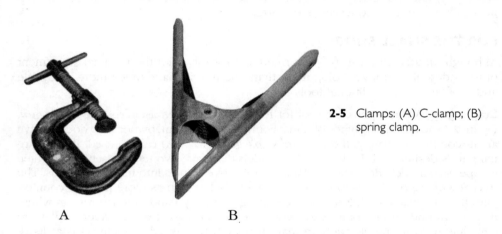

2-5   Clamps: (A) C-clamp; (B) spring clamp.

A   B

Hand drills   Making frame repairs frequently calls for installing or removing screws, drilling pilot holes, or drilling dowel holes. A *brace* or a *hand drill* is used for these tasks. The brace (FIG. 2-6A) is used for twist drills and auger bits having tapered square ends, such as the auger bit in the photo. The square tapered end fits into the notched, two-jaw chuck of the brace. The offset handle of the brace permits considerable rotational force for the turning operation. The manual *hand drill* (FIG. 2-6B) (and the more popular electric and pneumatic drills) has a three-jawed chuck, commonly called "Jacob's chucks." Three-jaw chucks are ideal for holding and turning twist drills, often called "drill bits," that have round or hexagonal shanks. There are several brands of three-jaw chucks; however, only one is the original Jacob's chuck. The manual hand drill (FIG. 2-6B), although considerably slower than the electric or pneumatic models, can do anything either power model can do.

Auger bits & twist drills   *Auger bits*, are used exclusively for drilling wood. A tapered lead screw draws the cutters into the wood during the drilling process, making the only real effort that of turning. Auger bits commonly range in size from $\frac{3}{16}$ to 1 inch, graduated in sixteenths. The number stamped, usually on the square shank of the bit, means sixteenths; for example, a bit stamped 9 would drill a hole $\frac{9}{16}$ inch in diameter. Twist drills, screwdriver bits, and other types of drivers can be purchased with square drives to fit the brace. Some multiblade screwdrivers, the kind that fit into a common handle, are excellent for use with a brace because they have small "ears" on the shank that prevent them from turning in the chuck.

Twist drills (FIG. 2-6B) are used to drill metal, plastic, or wood. They have no lead screw, so all penetrating and rotary force must be applied by the operator. Screwdriver bits can be purchased for the hand drill in round or hexagonal shanks, or they can be readily made from an extra screwdriver by cutting off the handle, an expensive way to get a screwdriver bit.

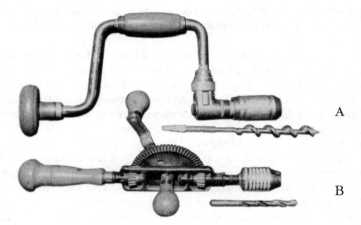

2-6 Manual hand drills: (A) brace with auger bit; (B) hand drill with twist drill.

Webbing stretchers If a unit has webbing that needs repairing or replacing, some kind of *webbing stretcher* is a must. A tool that can pull the full width of the webbing evenly is all that is necessary. Several types of webbing stretchers are shown in FIG. 2-7.

Lever. This type is for use with fabric webbing. It was first designed and built by the author in 1982 (FIG. 2-7A). The main advantages are that: (1) it can be used on webbing pieces no more than 1 inch longer than the frame it is to span; (2) it has direct lever action, permitting great leverage and stretching with minimal force required; and (3) because of the unique head design, it is capable of providing ample

2-7 Webbing stretchers: (A) lever; (B) webbing plier; (C) blade; (D) offset.

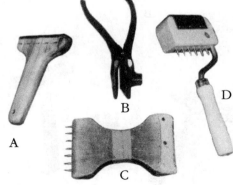

travel to tighten even long strands with no slippage and no marring of the wood surfaces it is placed against. Numerous users have indicated that this stretcher is the easiest and most comfortable to use of the four varieties pictured.

Webbing plier. This plier (FIG. 2-7B) is extremely versatile. It can handle fabric, plastic, rubber, and some metal webbing. (I have even used it to stretch polyester fabric over wooden screen-printing frames.) This tool can grasp any of these and provide plenty of leverage to tighten them. The major disadvantage is that the plier, all metal, can scratch or mar a frame easily. It thus requires a good padding between it and any show wood.

Offset. A comfortable stretcher to use for fabric webbing, it has excellent leverage and a nonslip, no-mar head which can be used against show wood frames with no danger of marring (FIG. 2-7D). Two minor drawbacks of this type: (1) it requires the operator almost to "thread" the webbing between the offset handle and the prongs, holding the handle in a rather awkward position; (2) a slightly longer strip is needed for grasping than with the lever stretcher.

Blade. Perhaps the granddaddy of the industry (FIG. 2-7C), it is used for fabric webbing and does an adequate job of stretching. It seems rather awkward to use, since it requires the operator to have the hand(s) between the webbing and the stretcher or work from the sides to achieve leverage. It does require at least 6 inches more webbing to reach the prongs than the outside dimension of the frame.

Any of these stretchers can be used on roll material, of virtually infinite length. If webbing is being used directly from the roll, the extra lengths needed for the offset and the blade units would be of no consequence. One other type of stretcher, the band stretcher (not shown); is used for installing metal webbing.

Sinuous spring puller These tools (FIG. 2-8) are extremely handy for installing sinuous springs. The springs can be installed without a puller, but when many pieces have to be installed or time means dollars, the time and strain saved through the use of the puller is worth many times the nominal cost of making one. Each is constructed simply enough for a do-it-yourselfer to make. The *lever puller* (FIG. 2-8B) is made with a piece of 1¾-inch-thick hardwood, a heavy coat hanger, or an ⅛-inch wire rod, a couple of pieces of rubber, and a ¼-inch machine screw and nut. The *band puller* (FIG. 2-8C) is made with a piece of ⅛- × ⅝-inch band iron welded to a ½-inch steel rod. The

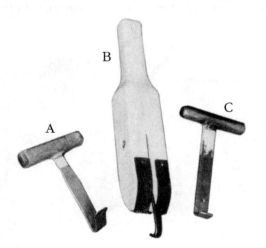

2-8 Sinuous-spring pullers (each shop-made): (A) light duty; (B) heavy duty; (C) lever.

sheet-metal puller (FIG. 2-8A) is made from a piece of 16-gauge stainless steel and a piece of ¾-× ¾-inch hardwood for a handle. All of them work well.

Mallets   *A mallet*, is an essential tool. Of the variety shown, two are most preferred: the rawhide (FIG. 2-9F) and the white hard rubber (FIG. 2-9E). The rawhide is preferred for its endurance, lightweight, nonmarring characteristics, solid face, and lack of splintering. The white rubber mallet is used extensively for installing tack strips; smoothing and contouring some slightly irregular padded edges, corners, and shoulders; and for fitting seat units into their metal pans (such as theater seats). The black rubber mallet, although practically identical to the white, leaves occasional black marks. All mallets used in upholstering feature nonmetallic faces.

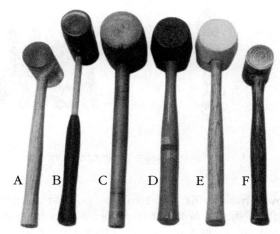

2-9   Mallets: (A) plastic head; (B) lead-filled head; (C) wood head; (D) black hard rubber; (E) white hard rubber; (F) rawhide.

Hand screw   Where pressure is needed for parallel or slightly angled surfaces, the *hand screw* (FIG. 2-10) is a very handy tool. Each handle has right- and left-hand threads on the rod, can turn independently of the other, and when worked singly or in unison can exert tremendous pressure. This tool is frequently used to clamp broken rails, uprights, slats, or other frame members while a good-quality wood glue is drying.

2-10   Hand screw: used for clamping parallel and near-parallel surfaces.

Bar clamps   When frame repairs require a squeeze on spans exceeding 10 inches, *bar clamps* (FIG. 2-11) are handy. Since not all frames require this type of repair, the bar clamp has not been identified as necessary for the do-it-yourselfer. The *pipe-base* *bar clamp* (FIG. 2-11A) has infinite adjustment of the stationary jaw, shown at the extreme right of the photo, while the *band-base clamp* (FIG. 2-11B) has fixed stops (notice the notch on the bottom of the bar just to the right of the jaw).

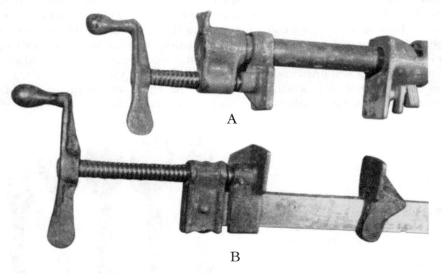

A

B

2-11 Bar clamps: (adjust the length of the bar); (A) pipe-bar clamp; (B) band-bar clamp.

Power hand drills The two most popular types of power hand drills for upholstering are shown in FIG. 2-12. Figure 2-12A is a pneumatic drill with a ½-inch chuck, chuck key, and a hex-shank, Phillips screwdriver bit. The electric drill (FIG. 2-12B) has a ⅜-inch chuck, chuck key, and a straight-slot screwdriver bit. Both drills are variable-speed and reversible. Both reverse and variable-speed control are essential for inserting and extracting screws.

A

B

2-12 Power hand drills: (A) pneumatic (with chuck key and hexagonal shank, Phillips screwdriver bit); (B) electric (with chuck key attached to power cord, and round shank, straight-slot screwdriver bit).

Stuffing regulator Occasionally some lumps, ripples, dips, or blobs are observed after the cover has been attached. The *stuffing regulator* is used to move small portions of padding to smooth the contour and at times fill some corner voids. Figure 2-13 shows the more common regulators: A—common *ice pick* (the least expensive); B—the officially labeled *Stuffing Regulator* (and ironically, the least useful); C—*"fancy" ice pick*. It should be noted that two are ice picks. Most upholsterers prefer the positive control afforded with the wooden handle. The ice picks have very sharp points, are very inexpensive (except for the fancy or "commercial" model, and the smooth, slim taper of the shank is precisely what is needed to give minimal disturbance of the cover fabric.

2-13 Stuffing regulators: (A) standard ice pick; (B) upholsterer's stuffing regulator; (C) household ice pick (with scalloped flange filed off the metal cap).

A   B   C

Needles If any deep sewing (through a few inches of padding), tufting, or buttoning is to be done, long, straight needles (FIG. 2-14) are essential. For taking long running stitches, the *single-point needle* (FIG. 2-14C) is generally used. Where stitches are taken repeatedly through cover and padding, as when sewing the cushion retaining groove, the *double-point needle* (FIGS. 2-14B and D) is used. Use of two points saves time by eliminating the need to reverse the needle direction each time it is inserted through the sewing materials. For installing buttons or for tufting, a *button-tufting needle*, with a handle on one end (FIG. 2-14A), is best. Many operators have put a "handle" on the second point of the double-point needle, creating a personalized tufting needle safe and comfortable to use. The tufting needle shown in FIG. 2-14A was made that way.

Welt board More often than not, the *welt board* (FIG. 2-15) is homemade. All that is necessary is a smooth, straight piece of close-grained hardwood (alder, poplar, maple, birch, cherry) that is 1½ inches wide and of a length convenient and sufficient, usually ranging from 60 to 80 inches. It is extremely useful for marking welt strips preparatory to cutting them from the cover fabric.

Trestle A *trestle*, or *horse*, (FIG. 2-16) is a tool of the trade. Many beginners, home-owners, and hobbyists, however, have done a lot of reupholstery working on the

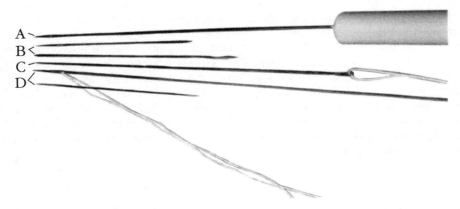

2-14 Needles: (A) tufting needle (converted from a round double-point; (B) diamond, double-pointed (eyes toward bottom); (C) round, single-pointed (eye at top); (D) round, double-pointed (eyes near bottom).

2-15 Welt board: made from alder wood ¾ × 1½ × 80 (length and thickness are optional).

2-16 Trestle: features recessed, carpeted top (a must); center shelf with raised sides and ends (very helpful).

floor, benches, or tables. As one gains experience working on trestles, it soon becomes apparent that the furniture is much more accessible with the trestle than with any of the alternatives mentioned. An upholsterer's trestle is always padded with a piece of carpet, heavy nylon, or vinyl fabric. This is to protect show wood from getting scratched and protect cover materials that have been installed on the unit from being snagged by wood splinters.

FOR THE PROFESSIONAL

Professional shops, doing a wide variety of jobs as easily and rapidly as possible, will have, in addition to the previously mentioned tools, most of those listed in this section.

Hog-ring pliers *Hog-ring pliers* (FIG. 2-17) are used to install hog rings. These fasteners are used extensively for fastening (1) two or more wires, such as edge wires and springs; (2) edge rolls to edge wires; and (3) cover materials to the metal and wire frame elements, as in auto upholstering and some types of office furniture. The most popular models are spring-loaded (FIG. 2-17B and D). The spring creates sufficient closing pressure on the plier to retain the hog ring within the slotted jaws, with no additional pressure from the operator; this feature is especially useful when the plier must be inserted into tight places to attach the ring. Without the spring tension (FIG. 2-17A), the ring often falls out of the jaws before it can be located and the crimping pressure applied. The spring-loaded and standard models come with angled or straight heads. For the auto upholsterer or the shop where a lot of hog-ring work is done, the *pneumatic hog-ring gun* (FIG. 2-17C) is a necessity (there are several other models of this gun besides the one shown).

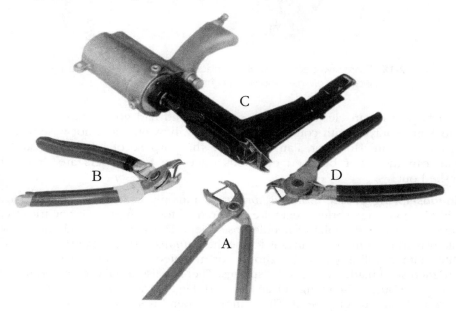

2-17 Hog-ring pliers: (A) inexpensive model, not spring-loaded; (B & D) spring-loaded, angled head (B), straight head (D); (C) pneumatic gun (carries a clip of 50 hog rings).

Foam saw This tool (FIG. 2-18) is so time-saving and does such smooth work that no commercial shop is without one. They are a bit too expensive for most hobbyists. It is more economical to pay a commercial shop to do the cutting. These saws can cut through foam rubber the full depth of the dual, counterreciprocating blades, which range from 6 to 18 inches in cutting length, with great ease and very smoothly. An occasional coating of silicone "dry" lubricant (WD-40 is one type) along the length of the blades and on the base plate greatly increases the life of the blades and increases the ease of passing the saw through the foam.

2-18 Foam saw: electric with 8-inch blade, sufficient for most needs (5-inch cushion foam shown on right).

I have used an electric knife to cut foam and had good success. Once the blades are sprayed with commercial lubricants, however, it is not wise to use it on foods again. However, Pam will serve the purpose of lubrication, and the knife can thereafter be used on foods once the foam fibers are thoroughly washed out.

Sinuous-spring cutter *Sinuous springs* (commonly called *no-sag springs*) are rather heavy-gauge spring-steel wire and are real tooth chippers if one tries to cut them with dikes, slip-joint or needle-nose pliers. The safe, easy, and only reasonable way to cut sinuous springs is to use the *sinuous-spring cutter* (FIG. 2-19). This cutter, either wall-mounted (as shown) or mounted on a table or bench, has a hardened steel blade that cuts the springs. There is also a built-in end crimper for turning the outward-curving cut end inward. Figure 2-20 shows the crimper readied to make the reverse bend. The completed bend is shown in FIG. 2-21. This reverse bending is necessary to prevent the spring end from (1) gouging into the frame, (2) tearing burlap and stuffing materials, and (3) squeaking as it rubs against the frame.

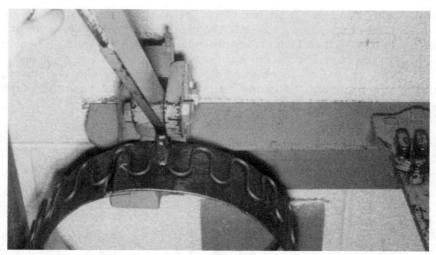

2-19 Sinuous-spring cutter with spring in cutting position.

2-20 Sinuous-spring crimper with spring ready to crimp back projecting end.

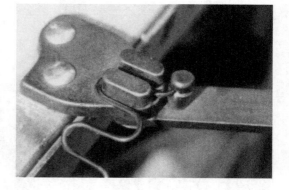

2-21 Sinuous-spring crimper with spring end crimped.

Sinuous-spring benders These hand tools are used in pairs to adjust bends and arcs in standard sinuous-spring material. The benders are placed over adjacent loops of the spring (FIG. 2-22) and with counterrotating pressure, the spring is bent as desired in either direction. Used in this manner, straight bends can be made neatly and easily. Without the benders, twisting and nicking of the spring wires is common, and bending is very frustrating. V arcs and Z arcs are easy to make. Figure 2-23 illustrates how the bend for a V arc is completed.

2-22 Sinuous-spring benders: a short spring section is beneath handles, ready to be bent into a V arc.

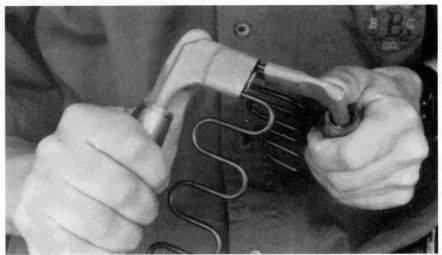

2-23 Sinuous-spring benders: at conclusion of acute angle bend for V arc.

Baker clip plier In all work with soft-edge seat and back suspensions, an edge wire borders the spring units. Where two wires are to be joined side by side, the Baker clip is used, and the Baker clip plier is used to clinch the clip around the two wires. Figure 2-24 shows the Baker clip plier with a baker clip in the jaws and another one alongside. You can see how closing the plier jaws will fold the three ears of the clip around the wires in one action. (This work could be done with a pair of slip-joint pliers but only when the Baker plier is broken or missing.)

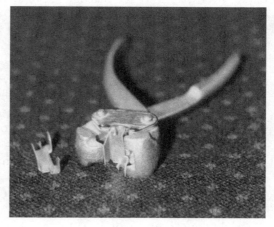

2-24 Baker clip plier with one clip loaded into plier jaws and a second clip alongside.

Button machine Most commercial shops and upholstery schools will have a *button machine*. The single-head unit shown in FIG. 2-25 is probably the most popular and cheapest of the button machines. It is a small, hand-operated, portable unit capable of making professional-quality buttons with ease. Other models are faster but will not do a better job than these well-designed hand models.

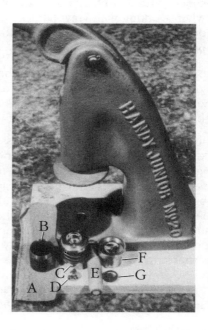

2-25 Button-making machine: (A) cover fabric; (B) fabric cutting die; (C) button-base retaining die; (D) button eye base; (E) wooden plunger; (F) cap retaining die; (G) button cap.

Shown at the base of the machine is a strip of fabric (A), a cutting die (B), button dies (C & F), wooden plunger (homemade, E), button cap (G) and button base (D). These components comprise everything necessary to make a fabric-covered button. One other accessory that is very helpful is a *spacer* or *cutting board* (a piece of hardwood or fiber material, like Masonite) to reduce the space between the upper and lower "plates" of the machine. Several thicknesses or folds of fabric can be placed on top of this spacer, the cutting die (FIG. 2-25B) placed on top and the handle activated, forcing the die down through the stacked fabric, cutting several "button covers" at one time.

To make a button,

1. Cut the "button covers" as indicated above or by driving the cutting die through single layers of fabric with a mallet.

2. Insert button base (prong or eye; FIG. 2-25D shows an eye base) into the button-base retaining die (C) with the prong or eye downward. Place the protruding shaft of this unit into the hole in the base of the press.

3. Place one button cover, the outside or finish side of the fabric down, into the cap retaining die (F).

4. Place a button cap (G) over the button cover so the convex face of the cap matches the concave center of the die.

5. Hold the cap retaining die (F) in one hand and with the wooden plunger force the button cap and cover down until it stops. This usually takes considerable pressure, especially if the fabric is heavy.

6. Place filled cap retaining die on top of the base retaining die and pull the handle down until it stops solidly.

7. Snap the handle down firmly two or three times to make sure the cap has clinched firmly to the base. *Caution:* This step may need to be modified when working with vinyl fabrics; much lighter pressure may be necessary, or the vinyl material will tear around the edges.

8. Remove the covered button and repeat for as many buttons as desired.

Another model of button machine; shown in FIG. 2-26. Has a pivoting base that speeds up the button-making process by as much as 50 percent. The left side of the machine has the cap die (a button cap is sitting on the edge); the right side is the button-base die (an eye base is sitting on the side of the die).

Klinch-it tool & clips Another indispensable tool for the professional is a Klinch-it tool (FIG. 2-27). This tool quickly and conveniently anchors coil springs to a fabric webbing base. It eliminates the need to fasten the springs by the outdated hand-sewn method. The tool is shown with strips of the clips on either side. A close-up of the business end of the tool (FIG. 2-28B) shows how the four prongs of the end clip are spread or flattened sideways as the tool is operated. Figure 2-28 also shows the bottom view of the carrier strip (A) and a top view of a strip of clips (C). The sharp points (four per clip) angle slightly outward in their normal position in the fabric carrier strip. When the tool handle is compressed, the crowned cap of one clip is forced downward in the center while the outer edges are retained by the tool tracks. This action forces the points outward, as shown with the bottommost clip in FIG. 2-28B. Had it been completely "clinched," the clip would have been torn from

2-26 Two-headed button machine: (increases production speed by as much as 50%).

2-27 Klinch-it tool and staples.

the remainder of the strip automatically, permitting the "clinched" clip to separate from the strip.

Blind tufting needle & clip This tool and its specially designed clip (FIGS. 2-29 and 2-30) permits installation of eyed buttons to completely finished furniture without stripping any of the cover panels. The work is done entirely from the front side. The close-up view (FIG. 2-30) shows the tufting needle in its protective cork retainer (A), fully loaded with a clip and tufting twine (B), and the clip showing the holes for the twine (center) and the needle hook (squared hole, top, C). The dime is included for size comparison.

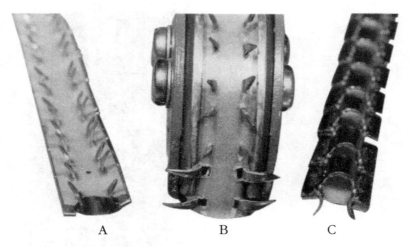

A   B   C

2-28 Klinch-it tool and staples, close-up: (A) bottom view of staple strip, showing partially curved teeth; (B) Klinch-it tool with bottom staple activated, showing how it is spread to "clinch" into cloth webbing; (C) top view of staple strip showing crowned top that forms over spring wires.

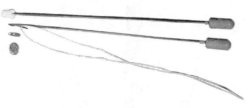

2-29 Blind tufting needles: shown in cork-tip protector (left); clip attached to needle, loaded with twine (center); a clip to the right (dime is for size comparison).

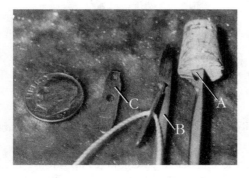

2-30 Blind tufting needles, close-up: details of the clip and the needle hook.

Gimp gun This is one of the specialty tools of the professional (FIG. 2-31A). This gun is used for attaching ends and folded corners of *gimp* to furniture calling for the decorative trim. The staples are almost totally invisible when applied because of their narrower heads. Figure 2-31 shows the difference in staple size between the pneumatic gimp gun (A) and the standard pneumatic model (B).

2-31 Comparison of pneumatic staple guns: (A) gimp gun with sleeve of gimp staples; (B) standard staple gun with sleeve of staples. Note the width difference.

A   B

Band clamp *The band clamp*, or *strap clamp* as it is sometimes called, (FIG. 2-32), a rather unique tool, is capable of exerting even pressure on all points of round objects, at all corners of square or rectangular items, and at almost any point along the convex curves or corners on irregularly shaped pieces. The strap is of considerable length, commonly ranging from 12 to 20 feet. It can girth rather large objects, even couches. The corner brackets are provided to protect sharp wooden corners from the abrading pressure of the tightening strap. The seemingly small wrench, resting on the adjusting nut, provides sufficient leverage to clamp most joints tight enough for quality glue joints.

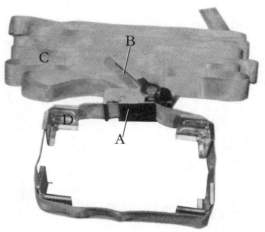

2-32 Band clamp: (A) ratchet head; (B) wrench; (C) nylon webbing band; (D) metal corner-protector brackets.

Fabric saw The *fabric saw* (FIG. 2-33), another specialty tool, is used by shops and schools doing production work, cutting out numerous identical panels of fabric (up to 250 at a time) with accuracy and relative ease. Custom shops and most schools have no use for such a tool, but where hundreds or thousands of the same panels are needed, the time and blisters saved over hand-cutting each piece soon pays for these relatively expensive machines.

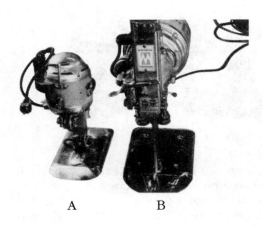

A   B

2-33 Fabric saws: (A) rotary or disk saw; (B) reciprocating double-blade saw. Silver State Suppliers

3
Furniture
styles

The majority of American households in the late twentieth century are selecting upholstered furniture for function and personal preference, involving a blending of contours and features rather than a strict period motif such as Mediterranean, French Provincial, Victorian, Duncan Phyfe, Chippendale, Queen Anne, Louis XVI, Empire, Adam Brothers, or Regency. For that reason, the discussion will focus on properties and shapes rather than historical or period features. If you wish a more specific exposure to historical styles, check out *Modern Upholstering Methods*, Tierney (1965), and *Upholstery*, Brumbaugh (1983).

An interesting experience awaits the visitor to almost any quality furniture store. The presentation by the experienced salesperson or interior decorator will center on "traditional" and "contemporary" styles because of the "American eclecticism" of furniture preferences. If a style carries a rather pronounced historical or period characteristic, it is called traditional. If the style is a combination of various period features (and most modern upholstered units fall into this category), it will be classified as contemporary, which simply means what people want now."

BASIC SEATING DIMENSIONS

When it is time to pad a piece of furniture, having a general reference for basic dimensions and what the resultant "feel" will be goes a long way to building self-confidence. For that reason, the following basics are provided—"basic" because there is such a variation in upper-body measurements (buttocks to shoulder) even between persons of the same height.

Backs

Four categories for back heights are presented. The measurement range is the distance from the top of the cushion (with no one sitting on it) to the top of the back, following the curvature of the face of the back, not the vertical distance.

High back (25 inches plus) *High backs* provide sufficient elevation from the seat to give a headrest for most adults. A back extending 25 inches or more above an unoccupied cushion can be classified as a "recliner back" because it is high enough for adults to lean back and have support for the whole upper body without any slouching.

Medium-high back (21–25 inches) Backs ranging from 21 to 25 inches will provide back support for most adults to about the tops of the shoulders. For people 5'-2" or shorter, the medium-high back will often provide a comfortable headrest as well.

Medium back (17–21 inches) If back support to about the shoulder blade is desired, the medium back will be sufficient for the majority of American adults.

Low back (14–17 inches) The low back is especially appropriate in conversation areas. A back extending between 14 and 17 inches above the cushion will generally allow an adult to turn partially sideways and rest an arm on the back of the unit without pronounced discomfort. For that reason, the low back is often called the "conversation back."

Seat depth

The "average" distance from the front edge of the cushion to the front of the back ranges from 19½ to 21 inches. Within this range most adults can sit with comfort. However, if a unit is being tailored for occupants with long legs and it is primarily for their comfort, longer dimensions (22–24 inches) will provide greater comfort. For those who may be built a little closer to the ground, a shorter seat depth (17–19 inches) would probably provide the most comfort, even to permit their knees to bend naturally at the front edge of the cushion while sitting normally in the seat.

Seat height

The majority of upholstered furniture has an unoccupied seat height, measuring from the floor to the top edge of the unoccupied cushion, between 15 and 17 inches. Balloon cushions and bench seats (not bench cushions) are two notable exceptions. Balloon cushions, because they are so much softer and compress so much more, are as much as 3 inches higher. Bench seats, because they don't compress as much and are designed more for short sitting periods, may be as much as 3 inches lower. Moving away from the unoccupied status, it is really the occupied or compressed height that has meaning.

Soft lounge height For the "sitting into" feeling, a unit with an occupied (compressed) seat height of 11 inches or less will do the trick. A seating height in this range is one from which senior citizens appreciate a helping hand to rise. This range gives that nice, soft lounge comfort.

Comfort lounge height The majority of upholstered furniture falls into this category. A compressed height between 11 and 14 inches is what most American furniture is designed to provide. This range provides that "expected" lounge comfort. Living-room furniture normally falls into this range.

Temporary lounge height A seat having a compressed height between 14 and 18 inches is comfortable only for moderate sitting periods. Dinette chairs and office furniture generally fall into this category.

Bench height Seats that have a compressed height of 18 inches or more fall into this category. For most of us, this seating height is meant for serious application and not for lounging comfort. Piano and organ benches and other chairs or benches that will require people to keep their backs straight fit into this group.

Arm height

The popular range for arm heights (the distance from the top of a compressed cushion to the top of a compressed arm rest) is between 8 and 10 inches. Distances greater than 10 inches elevate the shoulder abnormally high, giving the sensation of sitting sideways on a steep hill (if you use the arm rest). Arm heights less than 8 inches seem to increase tension by dropping the shoulder abnormally low.

STYLING FEATURES

We now turn to the styling of upholstered furniture. In the discussion that follows, terms that are more descriptive of shape and function will be used rather than historical or designer names.

Bench seat The *bench seat* is simply an extra-wide cushion that has no distinct divisions—divisions normally created by physically separate entities (two or more cushions, for example), welts, or seams that are intended to be obvious. Figure 3-1 is an example of a bench seat. This unit has a bordered pattern and features a pillow back. The arrows point to inconspicuous seams that are outside the striped "borders" of the basic fabric pattern. This approach permits adding sufficient material to both ends to permit the very wide pattern to be centered on the unit, an especially good approach for fabrics having extra-large floral groups or bordered patterns.

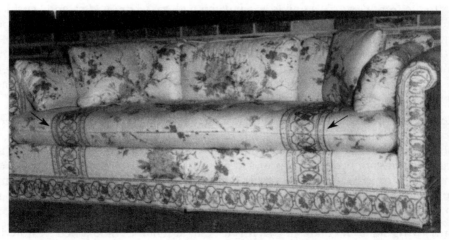

3-1 Couch with bench cushion and bordered pattern (arrows point to sewn seams).

Jointed seat panel One method used to break up long spans of solid fabric is to join panels at planned locations such as a centerline (FIG. 3-2) or where cushions meet (or both).

3-2 Jointed (seamed) seat panel at center of couch, finished off with flat panel skirt.

SEAT BANDS & SKIRTS

Narrow band Several approaches are commonly used to give variety in seat styling. Figure 3-3 features a full-width couch with extra-thick cushions (the balloon cushion), a rather wide seat panel, and a narrow band along the bottom front. The narrow band along the bottom of the seat is used to break up and add an air of lightness to the front view. This unit is accentuated with two extra-wide L and J cushions instead of a third rectangular cushion in the center. This design is especially appropriate because of the extra thickness of the cushions.

The arms are also formed with the "balloon" effect, sporting an almost oval frontal panel. Use of the extra-thick soft seat cushions permits use of shorter legs. The back is also padded extra thick, made in two sections, and buttoned to add character and break up what would otherwise appear as a very puffy back.

Matching band & seat panel Figure 3-4 shows a unit having extra-thick (5-inch) box cushions on the seat; contrasting, reversible, soft-stuffed, knife-edged, throw cushions on the back; with a matching (same width) band and seat panel along the front. The arms are padded and rounded with a pronounced plumpness. The inside-arm panel is made of one piece of fabric that covers around the front of the stump, is rounded over the top with tapering outside tucks, and finishes along the top to retain the parallel effect of the striped pattern.

Contrasting skirt A contrasting skirt is used to accent the "hide-a-bed" featured in FIG. 3-5, providing another finishing approach to a "contemporary" design. Double-buttoned, full knife-edge cushions are finished off with double-tuck corners on the front and a sewn-in square corner on the rear. The roll arms are wrapped completely around, featuring outside-tapering tucks at the top front of the arm.

3-3 Balloon seat cushions and arms, narrow seat band.

3-4 Wrapped arm, matching seat band, and box cushions.

3-5 Contrasting flat panel skirt, double-buttoned cushions (buttons on both sides of cushions), double-tuck corners on full knife-edge cushions.

Broad-faced seat panel Another way to finish off a seat section is the use of the broad-faced seat panel shown in FIG. 3-6, which blends in the balloon effect. The balloon motif is continued in the balloon arms (that again feature the outside tapering tucks to round the front top) and a crowned back accented with wide channels formed in 2-inch-thick soft (14 IFD, see Glossary and chapter 5) foam. The rounding

3-6 Wrapped arm; balloon waterfall cushions; and crowned channeled back with broad-faced seat band.

edges of the cushions can be created through the use of a dacron-wrap, extra-thick foam and tucking around the corners. (If the corners of the cushion panel are not tucked, the edges will be basically perpendicular with minimal rounding.)

Figures 3-7 and 3-8 show other applications of the broad-face seat panel, giving a feeling of mass and strength to the units. Balloon waterfall cushions are a common complement to the broad-faced seat. Figure 3-8 breaks the balloon effect by lightly padding a level arm and adding an unaccented (without welt) arm panel that covers the entire front of the arm.

3-7 Broad-faced seat band and balloon, waterfall seat cushions.

Still another style of the broad-faced seat panel is the full box frame topped with a double-layered, ultra-large biscuit-tufted (square pattern) cushion, as featured in FIG. 3-9. The "box" styling is maintained throughout by sewing seams along all edges. This style can be easily constructed to provide either more relaxing lounge comfort (by making the base of rubber webbing) or a casual bench seating (by topping the box frame with a plywood base).

ARM STYLES

Rolled arm Many variations have been given to the arms. The *rolled arm*, sometimes called *Lawson*, is often finished off with a welted panel covering the stump and welting around the front and top edges of the outside arm panels (FIG. 3-10).

The unit featured in FIG. 3-10, however, is finished off with the welted panel and no welt around the outside arm panels. Also featured is the typical welted box cushion and a formed seat, finished off with a welt around the bottom edge. The joint between the inside wing and inside arm (at the point of the arrow) is also finished with a welt.

3-8 Broad-faced seat band with lightly padded arm finished off on front with wide arm panels.

3-9 Ultralarge biscuit tufting on short-legged sectional.

3-10 Rolled or Lawson arm, wingback emphasizing welted union at bottom of wing panel. This union may be sewn to inner arm panel or fitted separately.

An alternative to the plain rolled arm is the lightly channeled rendition shown in FIG. 3-11. This style is pleasantly finished on the front through the use of a round-topped seat panel. A squared motif is accented with biscuit tufting on the knife-edged pillow back and finished off with the knife-edge seat cushion.

Flap panels The chair shown in FIG. 3-12 features a buttoned flap-panel arm, a buttoned knife-edged pillow back, and knife-edged seat cushion. The arm flap panel is blind-tacked along the top outside edge of the arm rail, then held down by stapling at the inside of the side seat rail through the use of a stretcher. The buttons can be of the double-prong or eye style and can be attached before affixing the flap to the chair or afterward. If the buttons are applied after the cushions are attached to the chair, the prongs or ties would extend through and be fastened to the arm base, of cardboard, webbing, or burlap.

A modification of the flap panel is shown in FIG. 3-13. Additional buttons are added along the top edges of the inside arm and inside back panels. The unit shown in FIG. 3-13 uses vertical channeling sewn along the lines of each set of buttons, with the top buttons attached near the outside edges of the frame. The balloon seat cushion is also buttoned. The arm stumps are finished with panels. Notice the heavily padded seat panel, which maintains the plumpness of this style.

3-11 Pulled biscuit tuft on back with lightly channeled arms and knife-edge T cushion.

3-12 Pulled diamond-tufted back and buttoned flap-panel arms.

3-13 Channeled, flap-panel, wraparound sandwich, buttoned at top and center of channeling.

Wrapped arm One of the modern variations in arm styling is the wrapped arm (FIGS. 3-4, 3-5, and 3-6). This is accomplished by wrapping a one-piece inner-arm panel around both the top and front of the arm without sewing any seams or adding welting. This style eliminates squared edges and panels to finish the arm stumps. Figure 3-4 displays the wrapped design on a straight arm; FIGS. 3-5 and 3-6 use the wrapping technique on a rolled style.

Button-pillow arm Pillow arms are made in much the same way as the pillow back components—a complete assembly of panels sewn together; then attached to the unit—with one basic difference: the arm is usually more heavily padded than the back prior to attaching the "pillow." Figure 3-14 shows a button-pillow arm accented with large welting of a contrasting color and material.

3-14 Buttoned pillow arms and back, accented with contrasting weld on cushions, and a box-fold skirt.

CONTRASTING WELT

The use of a contrasting color for welt has increased in popularity. Panels of plain contrasting colors also add a striking and pleasant accent to figured fabrics. Finishing both the outside arm and back panels in a contrasting color (as well as having contrasting welt) is popular.

BACKS

Tufting, channeling, and buttoning are the most popular ways of accentuating a back. These techniques are used with virtually every kind of fabric. It should be pointed out, however, that not all fabrics lend themselves equally well to all types of tufting and channeling. Some of these differences will be identified below.

Folded diamond tuft One of the favorite ways to finish off a back made of real leather or stretchable vinyl is by folding in the diamond tufts (FIG. 3-15) rather than sewing them in (soft-hand cloth fabrics are also occasionally finished this way). One reason why this style is more popular with leathers than other cover fabrics is that the folds stay in place better with leather. "Softer" fabrics seem to let the folds relax (sag) out of their intended location with wear.

3-15 Folded diamond tufting on back and seat.

Pulled diamond tuft Often, when the *effect* of tufting is desired, the back panel is cut oversize and when first installed, remains rather loose and almost floppy. Then buttons are added, *pulling* the loose fabric into three-dimensional contours of the desired design (biscuit, diamond, triangle, etc.). This style of tufting works best with the softer fabrics—those not having stiff or extra-tight weaves or stiff backing.

Sewn diamond tuft Where tufting that maintains an easily distinguishable tuft line yet gives a smooth contour is desired, sewing the design into the fabric is used. With this style, double-tapered tucks are sewn along each tuft line. The tucks taper to nothing at a radius the size of the button from the point of intersection of the tuft lines. This permits the buttons to carry part of the fabric into the padding, leaving the remainder outward with less "wrinkling" than with the pulled tufting. Figure 3-16 is an example of the *sewn diamond tuft.*

3-16 Sewn diamond tufting on crushed velvet back.

Pulled biscuit tuft Figure 3-17 is an illustration of *pulled biscuit tufting.* Note wrinkles primarily around the sides of the tufting, while the tuft lines are rather smoothly created.

Channels Numerous designs are created through the use of channels. Figure 3-18 utilizes contoured channels to accentuate the curved lines of the chair. The seat cushion was removed and placed in front to show how it was contoured to fit snugly into the irregular curves of the seat.

Camel back The *camel back* has one or two smooth flowing "humps" along the top. The unit shown in FIG. 3-19 illustrates one of the popular contours for the single hump. The double-humped camel back is seen with frequency on love seats (not shown).

Crowned back *Crowned backs* involve simply a single smooth upward arch along the back. Figure 3-20 is an example of this back style, featuring the crowned contour accented with what one manufacturer calls the "nail-head set trim" (described below) and lightly pulled tufting in extra-wide channels (used to simulate separate back cushions) to match the buttoned, reversible, knife-edge seat-cushion lines.

3-17 Pulled biscuit tufting on attached pillow back.

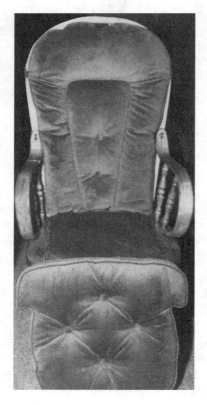

3-18 Contoured sewn channeling on back with double-buttoned seat cushion.

3-19 Camel-back sofa with rolled arms and center-seamed seat panel.

3-20 Crowned back with nail-head set trim and pulled deep channeling. Seat cushions are double-buttoned, knife-edge.

Nail-head set trim This feature could be classified as either a finishing technique or a back style, but because the sample presented (FIG. 3-20) deals more with the back than any other part of the unit, it is included with back styles. The cushions are attached low on the front of the top back rail; then a band is tacked on top of this, stuffed to the plumpness desired and either rolled over the top of the frame and attached to the rear of the top rail or stapled to the top edge with another band. Figure 3-21 gives a closer view of the decorative tack, band, and welt features.

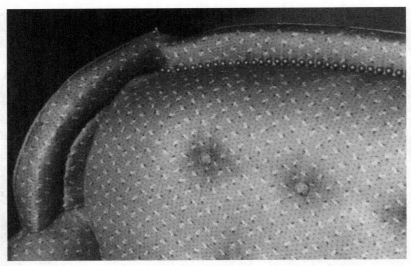

3-21 Close-up of nail-head set trim.

CUSHION STYLES

Squared crescent This cushion style (upper cushion in FIG. 3-22) can be made as a contoured toss pillow or a modified pillow cushion. The style shown is the modified pillow cushion. The lower cushion is attached so its tapered upper edge ends at about the point where the buttons of the top cushion are attached. One benefit of the modified pillow-cushion styling is that the attached panels reduce the "dust and crumb trap" by completely closing off the space between the cushion and the unit frame.

Flop cushion The idea behind the *flop cushion* is basically an enclosed pillow attached to a unit with an additional flap near the upper edge. Figure 3-23 shows the concept with a double set of flop cushions propped up to reveal the underside (the white lines in the centers are zippers). The reverse side of the cushion makes use of a less expensive fabric than that used on the cover. Figure 3-24 displays buttoned double flop cushions. Generally, the buttons are applied to the cushion only. As an alternative, the cushion can be buttoned to the foundation of the unit. Notice the multiple tucks at the rounded corner of the seat cushion. An interesting design is created with this approach.

Ram-horn cushion This cushion design is created by forming a tapering padding over the top of a back or the front of a solid seat so that the side view will appear the shape of a ram's horn (FIG. 3-25 shows this concept). To create this style, the cover is sewn to a backing panel to pull the cover around the edges to create the rounded appearance. At the seam of the two panels, an additional piece of fabric is sewn that will serve to attach the ends to the back post (in this case). A covered panel is used to finish off the ends of the back upright.

Clam-shell cushion An interesting approach that breaks the traditional cushion design is shown in FIG. 3-26. Either a deep tapered seam is sewn in one panel (in the same manner as sewn tufting), or two separate panels are sewn together. Gathers are taken in the center portions as the seam is sewn to create the clam-shell ap-

3-22 Squared-crescent upper-back cushions, gathered skirt and wings, accented with contrasting welt.

3-23 Double-flop-back cushions; single-flop arms; and eared, weltless knife-edge seat cushions.

3-24 Double-flop-back cushion; balloon, multiple-tuck seat cushion.

3-25 Ram-horn back cushion.

pearance. The center portion is drawn deeply into the stuffing by using a section of edge wire inserted into the sleeve created by the sewing, pulling and attaching it in the same manner as buttons.

Box cushions with welts   Figure 3-27A makes use of box cushions having welts. A contrasting color for the welt fabric can add an air of elegance and distinction. The contrasting welt is increasing in popularity.

3-26 Clam-shell back cushion.

3-27 Three popular seat-cushion styles: (A) welted box; (B) waterfall; (C) welted knife-edge.

Box cushions without welts A smoother feel is given to the standard box cushion by leaving off the welt.

Knife-edge cushions The *knife-edge cushion* (FIG. 3-27C) is quite a popular item. This basic cushion style is popular, and the inclusion of the contrasting welt is likewise increasing in popularity.

Knife-edge with double-tuck corners The double-tuck corner is used on knife-edge cushions when a rounded corner is desired rather than square. Figure 3-1 features a box-frame couch with a squared seat rail (not rounding into the arm stump) but sports a rounded double-tuck bench-style seat cushion. Figure 3-5 shows another use of the double-tuck of a narrower (distance between the tucks) styling.

Knife with multiple-tuck corners An alternative to the double-tuck corner is the multiple-tuck. The knife-edge seat cushion can use what might be classified as the "standard" multiple-tuck corner; the tucks are rather small and basically vertical. However, by increasing the width of each tuck a rather striking effect can be created.

Knife-edge with ears Where a casual or rustic atmosphere is desired, the addition of "ears" to the outboard side of knife-edge cushions is the answer. The lapping concept of the flop cushion is continued by the overlapping ears of the end seat cushions. Ears can also be added to the center cushion, overlapping the end cushions.

Waterfall cushions "Traditional" furnishings can be transformed into a somewhat more "contemporary" styling by replacing the standard box cushions with a *waterfall* variety. One of the more contemporary features added to the waterfall cushion is the ballooning characteristic (making the cushions of an extra-thick, softer (lower IFD) foam. Figure 3-27B uses a standard thickness.

COMBINED STYLES

As mentioned in the opening paragraphs of this chapter, most American upholstered furniture is a blending or combination of styling features. Weltless box cushions are blended with a seamed balloon arm, the flop-cushion back, and the broad-faced seat panel. Going back through this chapter, you can probably identify different styling features in many of the units shown. An attempt has been to open the avenue to contour and function as styling features rather than be locked into the traditional "period" classifications. If period styles are desired, it is suggested that the upholsterer review some of the books on historical furnishings.

FABRIC SELECTION

Fabric selection to portray any of the "styles" mentioned herein is strictly a matter of preference. Because of the modern attitude to have furniture possess looks and feel based on personal preference rather than a historical period or designer motif, the choice of fabrics has opened almost to infinity. What is desired by one person may cause another to shudder with a reaction to "poor taste." Some will say that a tapestry or matelassé pattern cannot be used on a contemporary piece of furniture. For the purist, that would be true, but to the contemporary homeowner or apartment dweller, there may be no period restrictions. Professional interior decorators will probably specify styles as they have been taught by artists rather than experienced tradespeople. And the reality is that not every interior decorator or upholsterer will agree on fabrics and styles. Put on what you like and what seems to go well with the rest of the furniture in the room. If you lack confidence to make that decision, consult with not one experienced professional but several. After getting several opinions, the choice becomes selecting among expert ideas.

4
Fabric

One of the questions asked most frequently of an upholsterer is "What is the best fabric to use?" The answer to this question will include a pronounced bias or personal preference, or additional information will be sought to determine "best for what?" No attempt will be made in this chapter to identify the best fabric in any general sense of the word. Rather, basic information will be presented from which you should be able to determine which is best for the particular application. We will first look at the primary types of fibers used in "general" upholstering.

TYPES OF FIBERS

Fibers used in upholstery cover fabrics are classified as either natural or synthetic. *Natural* fibers come directly from plants or animals. Among those coming directly from plants are cotton and jute, each being taken from plants bearing the same name. It could be said that one form of natural fiber is also made from metals (gold, silver, brass, aluminum), but these are not to be found in general usage; rather among the high-cost "special" fabrics. Of the various natural animal fibers, only two can be used in upholstery: silk (seldom to be found in modern commercial upholstery fabrics) and wool.

Synthetic, another term for polymer, is chemically speaking, really a "plastic." Of the scores of polymer bases, only five are commonly used in upholstery textiles: rayon, nylon, acrylic, polyester (Dacron), and polypropylene (Herculon or Olefin).

Each of the natural and synthetic fibers has properties that make it basically different from any other. As a result of these differences and the relatively few types of fibers acceptable for upholstery fabrics, it has been found beneficial to *blend* fiber types to obtain combinations of properties. With blending, there is really no limit to the combinations that can be made. But one principle seems to limit the multiplication of fiber combinations offered to the upholstery industry—volume sales. It is generally not the best economy to produce quantities of special fibers that have limited demand. Such "specials" would cost too much for the general public. Therefore, most textile production is limited to a relatively small number of high-volume "generals."

Textile technology has advanced so that a basic fiber composition can be modified to imitate others. For example, the popular natural upholstery fibers, cotton and wool, can now be treated and processed to look, feel, and react much like almost any of the synthetics. Neither, however, can be made to wear like several of them.

Synthetics can be processed to exhibit many characteristics of other synthetics as well as the natural fibers, even to the "itchy" feel of wool. Chemical changes in processing the fibers, changes in surface treatments, and mechanical treatments such as polishing, crimping, twisting, buffing, or roughing can alter significantly the final appearance and feel of a fiber, hence the fabric. A good chemist can create textiles that will confound even the experts. But the cost would be prohibitive for general textile production. For that reason, the majority of fabrics are made to take advantage of the basic chemical properties of each fiber.

BASIC FIBER PROPERTIES

The properties that seem of most concern to fibers have been condensed in TABLE 4-1. The information has been gleaned from two primary sources: (1) an internationally recognized polymer chemist and fiber specialist, and (2) several upholsterers having a combined experience in the trade of over 100 years. Although there is significant agreement between the two expert sources, complete consensus does not exist. TABLE 4-1 presents a condensed rating of the fibers. A clarifying discussion of each fabric follows.

Table 4-1 Property ratings on "popular" upholstery fibers.

Fiber type	Abrasion resistance* (Wear)	Flex life (90° bend, 1000 cycles)	Color retention (Fade resist.)	Fiber strength (Retention)	Feel (1 = smooth, 10 = harsh)	Look (1 = Shiny, 10 = dull)
Acrylic	5–10[a]	4–5	10[b]	10[b]	4	1–10[c]
Cotton	3	.3–.5	7–8	6	6	8[d]
Dacron (polyester)	7–8	4–5	9	9	1–10	1–8[e]
Herculon Olefin (poly-propylene)	9.5	17[f]	2	2–3	1	1[f]
Nylon**	10	20	3[g]	7[g]	1–10	3–10[g]
Rayon	1	.075	6	4–5	4	1–10[h]
Wool	4	20	1	1–2	10	10[h]

* Based on a scale of 1 to 10, 1 = low quality, 10 = high quality.
** This is the only fiber that has 100% elastic recovery.
a End wear (as with cut pile (velvets), is exceptional, side wear (looped weaves) is somewhat less.
b Has chemically built-in, ultra-violet screen, virtually unaffected by sun's rays.
c Can be processed to have highest luster of all fibers.
d Can be "polished" but loses "new look" rapidly. Sheen can be somewhat restored with surface treatments such as Scotch Guard.
e Has excellent "new look" retention.
f Flexing causes fibrillation (breaking into smaller diameter fibers which increases light refraction, giving the appearance of color loss and "frosting."
g Although ultra-violet light creates some frosting and yellowing of the polymer, nylon has been one of the most popular fibers for automotive upholstery.
h Can be produced with high shine but dulls rapidly with use.

Acrylic fibers Wear resistance across the fiber, as in flat and looped weaves (friezé is one example (see FIG. 4-1)), is relatively low, which makes it one of the least used fibers in twisted or braided weaves. End-wear resistance, however, is exceptionally high, making the acrylics especially suited for cut-pile fabrics (velvets and plush, FIGS. 4-2, 4-3, and 4-4). Acrylics have the highest resistance to ultraviolet light of all of the fibers. Since processing is relatively high in cost, acrylics are on the average more expensive than most other popular fibers. Another outstanding feature of this fiber is its high degree of softness, which imparts that luxurious feel to the velvets and plushes.

4-1 Frieze weave (the high-looped pile).

4-2 Plush (a long, delicate cut pile).

4-3 Velvet (a short cut pile).

4-4 Velvet (short cut pile with a pattern).

Cotton fibers Although previously rated quite low in wear resistance, the addition of surface treatments such as Scotchguard or Dow Gard significantly increases wear and cleanability. Without these treatments, cotton stains easily and cleans with difficulty. Cotton is especially suited for thin, delicate fabrics like chintzes (FIG. 4-5). Cottons have long been recognized for their "breathability," imparting a cool feel. Cotton fibers have not been processed (to date) to retain a new look for a long time. Because of this characteristic, cottons will be found in blends more frequently than alone. Polyester is one of the most popular blending fibers for cotton.

4-5 A chintz print (the smooth flat fabric).

Nylon fibers This fiber has come to be recognized as the workhorse of the fiber industry. Since it has the highest cross-fiber wear life of any material developed so far, it is used extensively in woven patterns for home and automotive applications. Nylon does not exhibit good end-wear characteristics yet. This explains why few nylon "velvets" are found. Recent modifications formulating have developed fibers that can impart an extremely soft feel that permits a wide range of textures. This is the only fiber that has full elastic recovery, which gives nylon fabrics a no-sag characteristic after prolonged or hard use.

Polyester fibers Dacron, by duPont, is the most well known brand name for polyester fibers. This fiber is recognized for a new-look retention for significantly longer periods than most other fibers. Polyesters also have excellent color and fiber-strength retention when exposed to sunlight. One recent formulation has imparted a superior softness to the fiber, making it a top candidate in automotive "plush" upholsteries (Rolls Royce plush velvet, for example). One drawback of polyester fibers is a rather low side-flex resistance; if the fabric experiences a lot of bending or waving, it does not wear well.

Polypropylene fibers Two very popular brand names for this fiber are Herculon and Olefin. These fibers exhibit very low ultraviolet resistance in color and strength retention. However, if the fabrics are kept from direct exposure to ultraviolet light, they are extremely strong and wear well. Polypropylene fibers take dyes well and are among the lowest-cost fibers to process. High strength and low cost are the major factors in its mass-production popularity. This fabric can be found on most lower-priced furniture.

Rayon fibers This fiber family has the lowest flex life of all fibers in use. It does not wear well in either end or cross-fiber applications. It "fuzzes" easier than others. However, it does have some advantages: brilliant colorability, excellent dye fastness, and soft feel. It is used in blends where high color contrasts are desired, making it suited for brocatelles (FIG. 4-6), damasks (FIG. 4-7), tapestries (FIG. 4-8) and matelasses (FIG. 4-9). It can be produced with high luster but will dull rapidly—a short-lived new look.

4-6 Brocatelle weave (the embossed one).

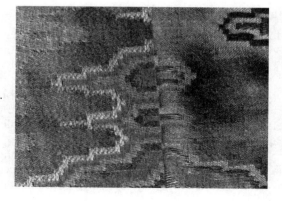

4-7 Damask (the reversible weave).

4-8 Tapestry (tight low loops with pronounced warp threads).

4-9 Matelassé (a double fabric woven together).

Wool fibers Normally rated low in cross-fiber wear resistance, wool exhibits very high end-wear resistance. It is second only to acrylic in cut-pile wearability. Wool possesses a natural look that is difficult to match by other fibers. This is due to the scales along the fibers, which not only impart a natural harshness, the itchy feel, but defy creation of a shiny or lustrous fiber. Because of the absence of shine and a natural air-pocket structure, wool has a warmth that is difficult to match. The major disadvantages of wool include a high allergenic property and relatively high price.

PATTERN ORIENTATION & NAP

Most upholstery fabrics are woven on looms. Threads running the length of the loom are the *warp* threads; those going across the loom are called *woof* (some call these *filler* threads). Patterns and nap are usually oriented along the warp. If a fabric has a floral pattern, for example, the flowers will be oriented to appear upward when looking from one end of the roll. If the fabric has a nap, it will be oriented so that it lies along the length of the fabric, in line with the warp threads. Some patterns will be oriented along the woof. Some will have the flowers sideways to the nap. This is especially true of newer velvets.

The orientation is very important to the upholsterer. Every panel to be placed on the unit must be oriented in the proper direction, or the unit will appear to be covered with fabric of different hues or the patterns will be going in different directions. Both conditions are avoided wherever possible. There will be exceptions that cannot be avoided. The general rule in the trade is *Flowers up, nap down*. This is easy to remember because flowers usually grow upward, and when people take a nap, they usually lie down. The arrows in FIG. 4-10 show the "down" direction for most of the common panels.

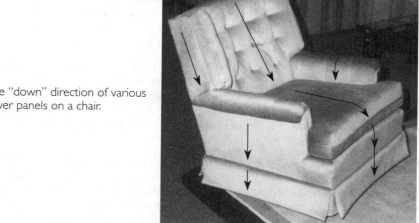

4-10 The "down" direction of various cover panels on a chair.

WEAVE TYPES

Numerous types of weaves are used in upholstery fabrics, but because of the large number and the relative similarity of many, only the more distinguishable of the types will be discussed here.

Brocatelle A fabric showing significant relief (third-dimension contouring) in the majority of the pattern, making this weave seem partly embossed. The terms *brocade* and *brocatelle* are occasionally used interchangeably; both have a distinct embossed property. Popular fibers are cotton, rayon, and nylon. The figure and background are usually tightly woven (see FIG. 4-6).

Chintz A thin, lightweight, flat-surfaced, printed fabric, usually with small, delicate designs. Cotton and rayon are the more popular fibers. This material comes polished or unpolished. The sample shown (FIG. 4-5) is polished (a slight sheen can be seen in the photo). This material is so much thinner and lighter than the majority of upholstery fabrics that some think the term *chintz* ("chintzy") was coined to signify inferiority or cheapness. That is not quite accurate, but is true that wear resistance of chintz is low compared to most other weaves.

Damask A flat, tightly woven fabric that unlike most others, is virtually reversible. The back of the fabric has the weave oriented exactly in reverse to the front and for all practical purposes, looks like the front. This can be observed on either side of the fold line in FIG. 4-7. Popular fibers include wool, cotton, rayon, and occasionally linen.

Frieze (pronounced "frizay") A woven, distinctly looped pile fabric, rather narrow loops extending upward, creating a springiness to the surface. This weave can have all the loops of the same level or can be figured with the background of a lower profile than the loops (FIG. 4-1). Popular fibers include wool and rayon, nylon being by far the most popular.

Matelassé (pronounced "mat-la-zay") A double or compound woven pattern that gives a double-relief characteristic where the pattern raises and the background appears to be negatively embossed. This weave is basically two separately woven layers interwoven at pattern locations. A section of the top layer has been cut out of the right side of FIG. 4-9 to show this characteristic. Cotton, nylon, and polyester are popular for this weave.

Plush A deep, thick, cut pile with a super-soft surface feel (FIG. 4-2). In the photo, the nap has been brushed in different directions to illustrate the pile depth and the significant change of light reflection and absorption; the pile is deeper and thus feels softer than that of "velvet." Wool has been used, with acrylic taking over the majority of this weave. Polyester has recently entered as a top-of-the-line plush fiber.

Velvet A short, thick cut-pile fabric (FIGS. 4-3 and 4-4) made from thin fibers of cotton, acrylic, rayon, nylon, and viscose, with polyester and polypropylene coming on line. The sample in FIG. 4-3 has also had the nap brushed in opposite directions to demonstrate the shortness of the pile (in comparison with the plush sample in FIG. 4-2) and the similar change in light reflection. Figure 4-4 is a patterned nylon velvet possessing a very high shine (this gives it a snowy appearance because of the white light reflections).

Tapestry A low-profile, very tight-looped weave. The individual threads are usually tightly twisted, giving a rather hard feel to the weave and a corresponding absence of any "fuzz" (FIG. 4-8). Wool and cotton are the prevalent fibers used in this weave.

ESTIMATING YARDAGE FOR COVER

There are three basic ways to determine the amount of fabric necessary to cover or recover any piece of upholstered furniture: yardage charts, pattern layout, and educated guessing (experience).

Yardage charts

Yardage charts have been used for decades by upholstery shops and schools to help customers and beginning upholsterers gain a feel for the amount of fabric necessary to cover a unit. These charts, published by fabric manufacturers, provide illustrations and yardage estimates for the more popular furniture styles. Figure 4-11 shows a chart put out by General Plastics, a division of General Tire and Rubber. Figure 4-12, produced by Uniroyal, is one of the more popular charts in current use. Notice that no dimensions are provided with the illustrations generally, so the user is left to interpret the illustration size in comparison to the unit. Under most circumstances,

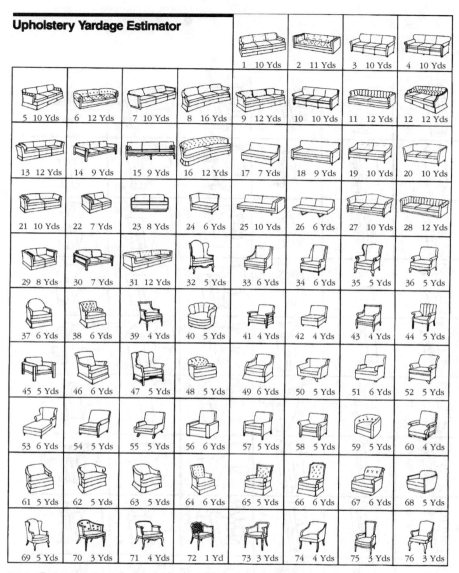

Upholstery Yardage Estimator

		1 10 Yds	2 11 Yds	3 10 Yds	4 10 Yds

5 10 Yds	6 12 Yds	7 10 Yds	8 16 Yds	9 12 Yds	10 10 Yds	11 12 Yds	12 12 Yds
13 12 Yds	14 9 Yds	15 9 Yds	16 12 Yds	17 7 Yds	18 9 Yds	19 10 Yds	20 10 Yds
21 10 Yds	22 7 Yds	23 8 Yds	24 6 Yds	25 10 Yds	26 6 Yds	27 10 Yds	28 12 Yds
29 8 Yds	30 7 Yds	31 12 Yds	32 5 Yds	33 6 Yds	34 6 Yds	35 5 Yds	36 5 Yds
37 6 Yds	38 6 Yds	39 4 Yds	40 5 Yds	41 4 Yds	42 4 Yds	43 4 Yds	44 5 Yds
45 5 Yds	46 6 Yds	47 5 Yds	48 5 Yds	49 6 Yds	50 5 Yds	51 6 Yds	52 5 Yds
53 6 Yds	54 5 Yds	55 5 Yds	56 6 Yds	57 5 Yds	58 5 Yds	59 5 Yds	60 4 Yds
61 5 Yds	62 5 Yds	63 5 Yds	64 6 Yds	65 5 Yds	66 6 Yds	67 6 Yds	68 5 Yds
69 5 Yds	70 3 Yds	71 4 Yds	72 1 Yd	73 3 Yds	74 4 Yds	75 3 Yds	76 3 Yds

4-11 Estimator chart from a U.S. vinyl fabric manufacturer. General Plastics

even the novice can select from the alternatives with reasonable reliability. One of the outstanding features of these charts is that the estimates are sufficiently generous to accommodate pattern repeats up to 15 inches when a pattern layout is used with care. These charts do not contain illustrations of all furniture styles, only some of the most popular. But long field usage has indicated that from those styles shown, close estimates can be guessed for many other styles.

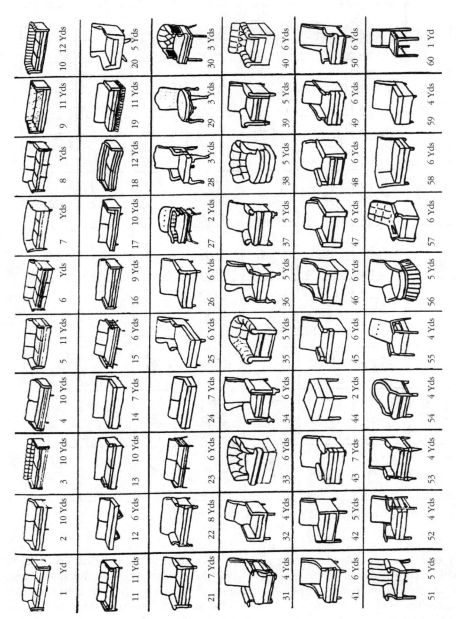

1 1 Yd	11 11 Yds	21 7 Yds	31 4 Yds	41 6 Yds	51 5 Yds
2 10 Yds	12 6 Yds	22 8 Yds	32 4 Yds	42 5 Yds	52 4 Yds
3 10 Yds	13 10 Yds	23 6 Yds	33 6 Yds	43 7 Yds	53 4 Yds
4 10 Yds	14 7 Yds	24 7 Yds	34 6 Yds	44 2 Yds	54 4 Yds
5 11 Yds	15 6 Yds	25 6 Yds	35 5 Yds	45 6 Yds	55 4 Yds
6 Yds	16 9 Yds	26 6 Yds	36 5 Yds	46 6 Yds	56 5 Yds
7 Yds	17 10 Yds	27 2 Yds	37 5 Yds	47 6 Yds	57 6 Yds
8 Yds	18 12 Yds	28 3 Yds	38 5 Yds	48 6 Yds	58 6 Yds
9 11 Yds	19 11 Yds	29 3 Yds	39 5 Yds	49 6 Yds	59 4 Yds
10 12 Yds	20 5 Yds	30 3 Yds	40 6 Yds	50 6 Yds	60 1 Yd

4-12 Yardage Chart. Uniroyal

Pattern layout

The *pattern layout* is the most accurate of the three estimating methods, the recommended method for the learner. Using the layout builds confidence to cut into that expensive fabric. It also builds confidence in the estimating charts by increasing an understanding of layout principles. The layout is especially useful to ensure that (1) every piece of cover is accounted for, (2) nap and pattern are oriented properly, and (3) there is sufficient material to do the job. It can be a sickening experience to approach the end of a job and find that there is not enough fabric to complete the project.

Avoid a headache. Use a pattern layout.

The pattern layout is an excellent teaching tool. A significant additional benefit of using a layout is that the inexperienced can confidently cut out every panel of cover, mark it on the reverse side, and stack it in the order to be installed on the unit. This can save significant time over cutting a panel and fitting it to the unit, then going back to cut another panel, fitting it, and repeating this process for each panel. (Professional upholsterers usually cut all panels at one time, stack them as indicated, and go to work.)

Two samples of pattern layouts (FIGS. 4-13 and 4-14) are included for reference. The sketches shown are sufficient and preferred. Just be sure that the major dimensions are included and accurately calculated to verify total yardage. Figure 4-13 is for a padded, high-back, open-armed rocking chair with no skirt. Figure 4-14 is for a couch measuring 67 inches between the arms with three reversible seat cushions and a plain, unpleated skirt all around.

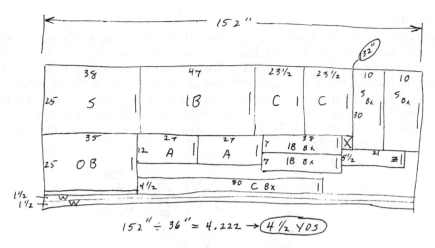

$$152'' \div 36'' = 4.222 \rightarrow \boxed{4\frac{1}{2}\ \text{YDS}}$$

4-13 Pattern layout (rough-sketched) for an open-arm chair.

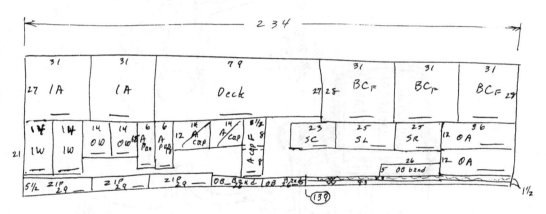

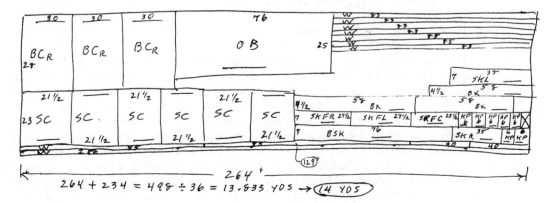

$$264 + 234 = 498 \div 36 = 13.833\ \text{YDS} \rightarrow \boxed{14\ \text{YDS}}$$

4-14 Pattern layout (sketch for a three-cushion couch with wings).

Neither of the layouts makes an allowance for pattern matching. If a fabric had a repeat of 27 inches, the yardage necessary might have to be increased by one-third to one-half.

The couch layout (FIG.4-14) minimum of 14 yards is unusually high. On this particular unit, the back panels for the back pillow cushions and the decking were all made from the cover fabric. This is done only when the fabric is cheaper than sewing labor or the customer prefers the matching fabric on all areas. (Some have suggested that the use of the cover fabric for all panels is a sign of higher quality. That is not necessarily an accurate assumption).

Any deviations from a standard flat covering (such as gathers, pleated skirt, deep channels, etc.) will increase the required yardage significantly.

The single short lines to the right of each panel in FIG. 4-13 and those at the bottom of each panel in FIG. 4-14 are *orientation marks*. These indicate the "bottom" or "down" direction of each panel. Putting these marks on the reverse side of the actual panels can save significant time when orienting and fitting them to the unit.

Be sure to use the maximum dimensions for each panel, and mark those dimensions on the layout as shown. From these overall dimensions, (1) a constant check is possible to verify that the panels that you're trying to get out of the width (generally 54 inches) is not being exceeded; (2) the total linear measurement can be quickly and accurately determined simply by adding up the dimensions along the edges.

Freehand sketch the layout, attempting to represent proportion rather than taking the time to make straight lines and accurate measurements. Figures 4-13 and 4-14 have been freehand sketched and reproduced in that form to depict the practical way to make a pattern layout and to show that meticulous care in drawing is not necessary. A representation is all that is necessary on the layout sketch. It is important that (1) all panels are accounted for, (2) all panels are oriented properly, and (3) there is sufficient fabric for the job.

Educated guess

When it comes to getting things done, there is no substitute for experience. Most experienced upholsterers can exercise the educated guess and estimate within one-fourth to one-third of a yard the fabric needed for almost any job. The main problems with this method are that (1) it takes considerable time to gain the experience necessary to estimate yardage for the different styles and sizes, and (2) it requires a good perception for differences and then a good memory to keep track of all the variations and their respective fabric requirements.

PATTERN MATCHING & PATTERN REPEAT

A picture of a heavy plaid fabric (FIG. 4-15), reveals what is meant by *pattern matching*. Notice that the same plaid pattern is matched from the inside back cushion to the seat cushions (even when interchanged or reversed), over the seat panel, and to the skirt. This is not really difficult; it just takes planning. Figure 4-16 shows that the inside arm panel is mismatched to the decking. Although the decking on this unit was of the cover fabric, no attempt was made to match the pattern in either direction (and this is quite acceptable for most work).

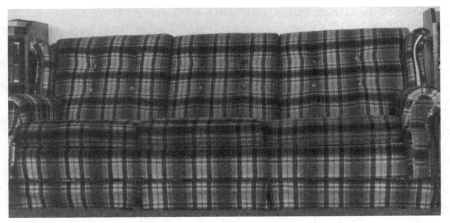

4-15 Pattern alignment (matching) from back down to skirt.

4-16 View of decking revealing mismatch of inside arm, inner back, and decking.

The view shown in FIG. 4-17 reveals a slight mismatch between the vertical pattern of the pillow back and the inside arm panel. Good-quality workmanship would definitely match these two components. If a match was desired, the match could be simplified by installing the inside back before the inside arms. Following this procedure, the smaller inside arm panels could be moved to match the pillow back and the decking with greater ease and less material loss than trying to match the back to the arms. To match the inner arm panels to the rest, they must be cut sufficiently oversize or laid out with that intent in the first place.

4-17 View showing very slight mismatch of inner back and seat cushion, significant mismatch of inside back and seat cushion.

The illustration shown in FIG. 4-18 illustrates the pattern repeat along the vertical and horizontal axes. Notice that the vertical repeat is 7½ inches and the horizontal repeat 6 inches. All fabrics that have a distinct pattern will have a definite repeat in both directions, with one exception. Some fabrics having a very pronounced pattern will have only one pattern across the entire width of the bolt. These will usually be bordered patterns (FIGS. 3-1, 3-2, 3-25).

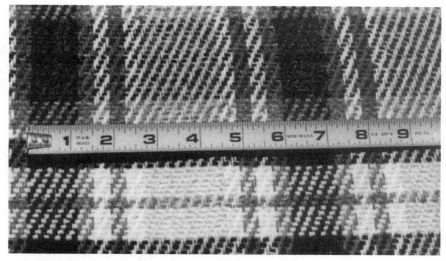

4-18 Pattern repeat, horizontal direction of unit.

5
Stuffing, padding, & muslin covers

This chapter deals with the preparations, characeristics of foams and paddings, and information on muslin covers.

CHARACTERISTICS OF STUFFING & PADDING

The most popular stuffing and padding materials in modern upholstery practices are cotton, Dacron, and polyurethane foam. Within each basic type are various compositions that will alter its characteristics, sometimes significantly. By layering or combining materials, a stuffing or padding can be tailored for almost any given application and will impart almost any firmness and resiliency. Yet with all the variations possible, it is important to recognize that all materials, no matter what their composition, will have a break-in and a break-down characteristic.

Some persons may slip in an unintentional declaration that a certain material does not break down. But perpetual longevity of resilient materials has not yet been invented. With this in mind, let us briefly identify what is meant by *break-in* and *break-down.*

The break-in period

The break-in period is a time when new stuffing and padding materials undergo changes in stiffness and cushioning characteristics—a period of stabilization. For example, resilient properties lose some springiness (as with cotton and dacron) or stiffness (as with foams).

Cotton will have slightly more resilient life during the first few months, then will settle down gradually and give five to eight years of rather consistent, firm support. Dacron and foam possess a distinguishable stiffness when new. Therefore, they feel firmer when first installed than a few weeks later. Because of these break-in characteristics, no new or newly reupholstered furniture possesses quite the same feel as it does a few weeks after it has been in use. So if that newly redone unit doesn't feel quite as you expected, give it a month or two. Give it a chance to break in, to stabilize.

The break-down period

This is the in-service lifespan of the padding material. When the "life" seems to be waning in the padding, it is going through the break-down period. All stuffing and padding materials will experience some form of break-down.

Cotton will eventually mat down to a rather firm, solid layer showing little life or springback. That matting process is greatly accelerated by moisture. Cotton does not decompose to lifeless powders, nor does it just seem to evaporate, unless rodents, mildew, or rot get into the act. Rather than a cell or fiber break-down, it seems just to compact.

The cotton used in upholstering is a composite material. The two major components of most cotton felts are known as *first-cut cotton linters* (which are the short, fuzzy fibers on the seeds) and *binders* (longer, coarser fibers of the cotton plant). When you purchase upholstery cottons, the quality and content are identified by two numbers, such as 70-30 or 85-15. The first number is the percentage content of cotton linters; the second, the percentage of binders. The higher the linter content, the softer and more lively the cotton felt. The higher the binder content, the heavier and stiffer the felt, and the quicker it will mat down.

Dacron tends to mat much the same as cotton. However, during the life of its resiliency, after the break-in period, it has a softer feel than cotton. Once Dacron mats down, you hardly notice that it is there. The word *Dacron* originates from the brand name duPont gave its line of polyester fibers; thus, if it carries the registration symbol (®), you know that it is manufactured by duPont. All Dacron referenced in clothing or upholstery is a polyester fiber, but not all polyesters are manufactured by duPont. The Dacron material used for upholstering is much like the quilt batting familiar to many. The most popular size for furniture upholsterers is 30 inches wide and 1 inch thick (in its relaxed form). A roll of 30-inch material will weigh between 13 and 15 pounds and contain between 39 and 45 lineal yards of material. It also comes in 72-inch-wide rolls weighing 31–36 pounds. This wider material is used primarily in the automotive upholstering business. Dacron has a very low compressive resistance, which gives it the soft feel. To roll the fibers between the fingers, however, will not give the impression of softness; they seem rather coarse and somewhat stiff. It is that stiffness that gives Dacron the "bounce" or spring for which it is known.

Dacron is generally used for the final surface padding of cushions, backs, and arms. It provides a very soft, almost "cushy" feel to the final product when the cover fabric is not pulled down tightly. It is also used to provide a resilient "filler" in corners of cushions, back-attached cushions, and other areas where a filler will not give the sensation of lumps. Often it is used as the filler for the low-profile channeled "sandwiches."

Foams undergo a cellular destruction as they break down. In this process, most foams will first mat, then decompose. Perhaps you have experienced a foam rubber that has "hardened?" That is the matting stage. This usually occurs at the front edges of cushions and points of major compression. The early stages of this hardening is not irreversible, however. Many cushions have been rejuvenated and their life span extended (see "Rejuvenation of foams"). In some foams, an embrittling and crusting follows the initial hardening. When this happens, the foam soon disintegrates into fine powder—the advanced stage of decomposition. From this stage, there is no known recovery, yet! Some foams display superior service life—the time span between installation and "back to dust." Currently, HR foams have proven so enduring that decomposition data have not been established—that's more than 12 years!

There is yet something better than HRs—*Ultracel*. To give you a brief idea of how good this foam is, we'll use a comparison of samples with a density of 1.8 cpf. Ultracel has about 25% higher resiliency than conventional foams, is the first high-support foam to achieve Compression Set resistance better than conventional foams, and still has a softer initial IFD and higher resiliency than conventional and HR foams. Hence, Ultracel wears longer, feels softer, and resists bottoming out longer than any other foam to date.

Understanding foams

Most, if not all, foams used in the upholstery industry today are polyurethane. Some people classify them as foam rubbers; others classify them as "plastics." Chemically, they are plastics. Behaviorally, they are rubbers. Polyurethane chemistry is extremely diverse and complex. There are thousands of formulations for making flexible urethane foam. That means that there are thousands of different foams. But only a few of those formulations are used extensively in general upholstery practices. TABLE 5-1 gives a condensed view of some popular foams; their major use areas, IFDs (indentation force deflections), densities (weight per volume), manufacturers, and the identification numbers from the various manufacturers.

IFD is a measure of the of pounds of pressure required to push a 50-square-inch disk (approximately 8 inches in diameter) 25 percent of the way into a foam specimen. Thus, if a piece of foam was 4 inches thick, the IFD is the pressure necessary to push the disk 1 inch into the foam. Upholsterers do not have testing equipment to identify IFDs, so, a trick of the trade: Figure 5-1 shows how a person, with a little practice, can get a close approximation of the IFD of a foam sample. Spread the fingers wide, and lean on the foam. This gives you a feel of the resistance the foam has to compression. If the piece of foam is big enough, use both hands, side by side, and lean on it. This gives an even more reliable basis for comparison (approximating the 50-square-inch disk). *Caution:* Do not try to identify IFD by pinching a piece of foam between the thumb and fingers. It is difficult even for experts to distinguish IFDs in this manner.

Density is identified by obtaining a foam sample that is 1 cubic foot in volume and weigh it. But if you don't happen to have a convenient cubic-foot size, find the weight (in pounds) of your piece, then multiply the length, width, and thickness (in inches). Divide that number by 1728 (the number of cubic inches in 1 cubic foot) to get the cubic feet you have. Now divide the weight of your sample by the number of cubic feet, and you have the density.

A few years ago, several foam manufacturers changed their numbering system to make it easier for customers to recognize, remember, and specify the IFD and density of their foams. TABLE 5-1 lists a few manufacturers and their identification systems. The two easier systems, Future and Foamex, basically use a five- or six-digit number. Future, in their five-digit number, lists the IFD by the first two digits, the last three representing the density (i.e., 14,090 = 14 [pounds for 25% deflection, the IFD] and 0.90 [density]). The density specification carries an implied decimal point after the first digit (090 = 0.90 pounds per cubic foot).

Foamex lists their foams in just the opposite order (i.e., 270,125 = 2.70 [density; remember the implied decimal point] and 125 [IFD]).

Crain's, R-21100-000 is a foam with 21 IFD and 1.00 density. The letters R and RA, and the terminating three digits -000 or -880 (from TABLE 5-1) are specifications of other characteristics, which will not be discussed in this book. There are many other properties that may be specified for particular foam applications (such as flame retardancy and compression set) that some professional upholsterers will need to know but are not within the need-to-know elements of this book.

Table 5-1 Urethane foams.

Use	Density	IFD @ 25%	Formula #	Manufacturer
Back foams	0.90	15	RA-15090-000	Crain
6 to 8 yrs.	0.90	14	14090	Future
	1.00	15	10015	Foamex
	1.00	13	R10	Ramco
12 yrs. +	1.80	10	HR 1810	Carpenter
	1.80	10	1810 (HR)	Foamex
Inexpensive	1.00	33	10033	Foamex
seat foams	1.05	33	RA-33105-880	Crain
1 to 3 yrs.	1.10	33	R22	Ramco
	1.12	39	R34	Ramco
Mid-range	1.40	31	R250	Ramco
seat foams	1.45	30	RA-30145-000	Crain
3 to 5 yrs.	1.45	36	RA-36145-000	Crain
Top-quality	1.80	24	18024	Foamex
seat foams	1.80	30	18030	Foamex
6 to 8 yrs.	1.80	35	18035	Foamex
	1.80	26	26180	Future
	1.80	50	50180	Future
12 yrs. +	2.60	28	26028 (HR)	Foamex
	2.60	40	26040 (HR)	Foamex
	3.25	50	32550 (HR)	Foamex
	2.70	27	27270 (HR)	Future
	3.10	33	33310 (HR)	Future
	2.40	50	50240 (HR)	Future
Church &	1.80	75	18075	Foamex
restaurant	1.80	70	70180	Future
Motorcycle seats	2.70	125	125270	Future
6 to 8 yrs.	2.70	125	270125	Foamex

For our purposes, only IFD and density must be considered together. The higher the density, the greater the mass (pounds per cubic foot or kilograms per cubic meter), the thicker the cell walls and the longer-wearing (higher quality) the foam. The higher the IFD, the stiffer the foam. Combined together, a high IFD with a low density indicates a stiff foam subject to short wear life—something useful for short-term padding, as for shipping. A low IFD and high density gives a soft foam that will endure longer—soft back padding for lobby furniture. Usually, the lower the density, the lower the cost of the foam.

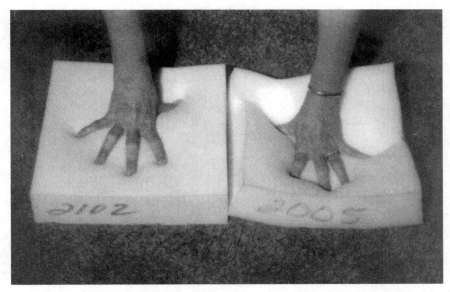

5-1 Differences in IFDs: the foam marked 2102 has an IFD of 50, density of 1.80; sample 2005 has an IFD of 12, density of 1.00.

HR foams The master of all upholstery foams is the HR or high resiliency foam. These foams have been formulated to give a much more lively (resilient) feel, a characteristic of building stiffness as they are compressed, and a much longer service life. Therefore, an HR foam with the same IFD as a standard *polyfoam* will have less tendency to bottom out, last longer, and seem more resilient during its lifespan. HR foams are typically high in density, ranging from 2.4 to 3.6. HR foams are more expensive than standard, about 30 percent. But when you consider the expense of time to do the job, the cost of cover fabrics, etc., in the upholstery or reupholstery process, the HR foam is by far the more economical buy.

The confusing way to order foams Some upholsterers may still use the "old" method to specify foam, with the terms *super soft, soft, medium, hard, extra hard*. However, a glance at TABLE 5-2 indicates that those terms are in need of clarification. For example, a medium *back* foam, with an IFD of 16, does not possess the same stiffness as a medium *bench* foam, where the IFD is 50. A medium bench foam installed in the back of an overstuffed chair would feel like you were leaning against a board. On the contrary, a medium back foam applied directly over a plywood bench seat would give the sensation of plopping through a pile of fluff and sitting on a board.

The proper way to order foams The appropriate way to order upholstery foam today is to specify both density and IFD. Consider the following in selecting a foam padding: (1) The back of a chair or sofa receives relatively little weight or stress. For this reason, it requires little stiffness to give a pleasant cushioning effect, and a rather low density will still be able to provide long life. (2) In church and restaurant seating, the

Table 5-2 Foam firmness ratings for different applications.

Use	IFD	Rating
Backs	≤10	Super soft
	12–14	Soft
	16–18	Medium
	20–24	Firm
	26+	Extra firm
Seats	20–24	Soft
	26–30	Medium
	32–36	Firm
	38–45	Extra firm
	46+	Super firm
Church &	40–48	Soft
Restaurant	50–70	Firm
	80+	Extra firm

benches get a lot of use. Usually the foam pad is placed directly over a solid wood or metal base. For those reasons, foams with a high density and high IFD are necessary.

Manufacturing and cutting foams   All polyurethane foams are manufactured in much the same way—a chemical reaction. Liquid chemicals are mixed intensely, then dispensed onto a moving conveyor belt with sidewalls. The conveyor belt and sidewalls are lined with a coated paper or a plastic film that moves at the same speed as the conveyor. This chemical reaction releases a carbon dioxide gas that causes a foaming, expanding action. Expansion can reach as high as 70 times the original liquid volume. A flexible, rubbery slab of foam is produced on a continuous basis. The foam slabs will be as much as 60 inches thick and range from 40 to 120 inches in width. The slabs are then cut into various lengths, called "buns," for curing and handling purposes. The cut bun length ranges from 100 to 200 feet in length. Figure 5-2 gives some idea of what the buns look like.

5-2 Polyurethane buns as they are placed to cure in a warehouse.

Rejuvenation of foams Sometimes a foam will seem to have good life yet display a section where the foam remains somewhat compressed. If hardening or granularization has not started, that section can be restored significantly. The extent of rejuvenation is dependent on the extent of decomposition. The process is very simple. Apply a jet of steam to the compressed area. This can be done with a commercial steamer, a steam iron, even a portable clothes steamer. As the steam contacts the cell walls, it creates a relaxation of the compressed rubber, permitting the cells to return to their original decompressed form.

STUFFING & PADDING

With frame prepared, the springs and burlap installed (chapter 13), the next step is applying the stuffing and padding. Almost any feel can be imparted through a combination of modern materials. For our first step, we shall look at the arm rail.

Many different feelings can be imparted to the arm. The fastest padding system is seldom the most desirable. For most furniture, built-up layers of cotton has been the rule. For a brief time after installation, this padding gives resiliency and comfort. But with time, the cotton mats down, and the result is a rather solid, lifeless armrest. Another fast way to pad an arm is to use a single layer of high-quality "standard" polyfoam. Both methods are frequently used in mass-produced furniture. But neither will give a resilient, comfortable armrest that does not bottom out. However, with the use of various combinations of foam; foam and cotton; or foam, cotton, and dacron; an infinite range of armrest properties can be created.

For an armrest that has body and life and seems not to bottom out, establish a firm, resilient base by any of the three following methods.

1. Use a layer of ¼ to ½ inch rather firm (50 to 90 IFD), moderate density (1.6 to 1.8) polyfoam.
2. Use the high-density (2.6 to 3.5) HR foams for this purpose. IFDs of 70 are about as high as the HR foams go.
3. Use bonded or standard (⅜–½ inch) carpet padding (new material, not used). This is my preference, and it becomes especially economical when you can use scraps left over from carpeting jobs, a natural in those shops that do both carpet installation and upholstery.

In all cases:

1. Cut the first layer of the foam, the "base," to the size and shape of the arm, allowing about 1⁄16 inch per side overhang. This breaks the hard, sharp edge of the wood frame.
2. Adhere with foam adhesive, or staple the base layer in place. If adhesive is used, spray both the arm rail and the underside of the foam pad with a good-quality foam adhesive. Allow the adhesive to dry until it has lost the wet look and feel; then apply in a rolling action. *Caution:* Try not to apply so much rolling pressure that the foam is stretched.
3. Finish off by applying top layers of foam and cotton.

A few tips for a couple of styles, to give an idea of how to go about this part of the job, follow.

Padding a flat inside arm

1. Apply foam adhesive to foam base and top of the arm rail, following steps 1 and 2 in the previous section.

2. Locate foam on arm and press in place. Begin at one end of the arm and lay the foam down in a rolling motion. *Caution:* Check the length before pressing the entire pad down. Even with care, it seems, the foam gets stretched or compressed a little. Figure 5-3 shows an operator "bridging" a piece of bonded foam so that it will not extend beyond the point intended. By seating the loose end in this manner, the remaining foam within the bridged portion compresses and adheres flat on the arm rail.

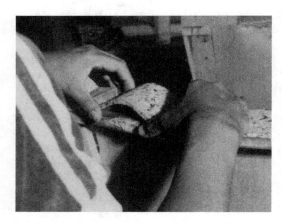

5-3 Using ¼-inch bonded carpet pad to create a "nonbottoming" armrest base.

3. Cover the base foam on the top of the arm rail and inside-arm area with the second layer of foam. The foam pictured in FIG. 5-4 is HR, with an IFD of 33 and a density of 3.10. (For the photo, one edge of the foam was glued to the outside of the arm rail, folded over the top, toward the inside of the chair, to locate where to make the cuts. The foam was then folded back toward the outside of the chair, the cuts made partially through just to show how the cuts are to be made, and left hanging down the outside of the arm. In normal installation, the dangling rectangles would have been cut out completely.) The first cut, to the far right, is made so the foam will snug against the back post. The second cut is made to leave enough material so that it can be pressed beneath the projection of the back post that extends over the arm rail. This ensures no open gaps in the finished arm.

4. Apply a layer of cotton padding over the foam. (Most quality upholstery will have cotton as the last layer in any padding. It feels cooler to the body than the foam alone.)

   ~Leave enough extra on the front to form a nicely rounded roll around the edges of the stump (FIG. 5-5). The rolled edge then forms a "trough" for the panel that will be added to finish off the front of the arm.

   ~Leave about 1½ to 2 inches extra at the bottom of the inside-arm and inside-back areas, which is then tucked under and between the arm, the back, and the seat to close any gaps. The extra cotton is shown lying on the deck and against the back (FIG. 5-6), ready to be tucked in.

~Staple the cotton at the tucked-under rounded edge so that no ripples or depressions are created. Notice how the operator is holding and stapling it in place in FIG. 5-5. The material around the top and right side will be formed and stapled in like manner. In FIG. 5-6, you can see that the left portion has been stapled into a nice roll. The top and inside portions are yet to be stapled in place.

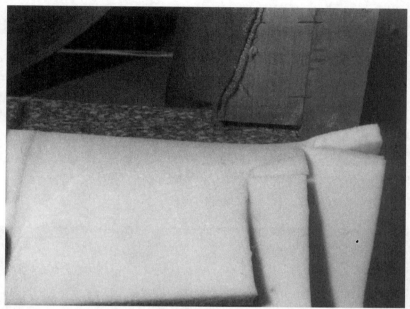

5-4 Contouring a layer of 1-inch foam padding for inside arm.

Padding a rounded or a Lawson arm rail

The Lawson arm (FIG. 5-6) may have its base established by any of the three methods above, then covered with a layer of 35–50 IFD, 1-inch foam, and topped with one or more layers of cotton. However, to achieve the rounding effect, the top of the arm must be built up in pyramid fashion, tapering toward the top. The final layer of foam in all cases should be one continuous piece, extending from the seat up and around the arm to the outer edge of the arm rail. The last layer of cotton should also be in one continual piece. Spray the underside of the foam, the surface of the arm rail, and the inside arm base with a foam adhesive; let dry past the wet stage; then press the foam in place. This holds it to the contour for application of the cotton and the cover. Cotton will normally stick to the foam so that it does not need to be glued.

MUSLIN COVER

Some of the higher-quality furniture is finished with a muslin cover. This cover is fitted and stapled just as the final cover would be, with two exceptions. The bottoms of

5-5 Stapling cotton to create a rounded or rolled edge at the front of the arm post.

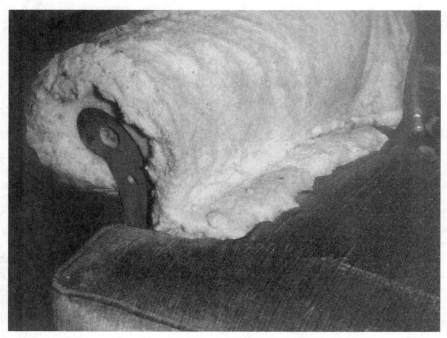

5-6 Lawson arm: cotton padding is final layer for inside arm. (Allow about 2-inch extra at bottom and back for tucking under.)

the inside-arm and inside-back panels *must* be stapled to the bottom arm and back rails, respectively, rather than the seat side and back rails. Installing a muslin cover can also serve as an excellent teaching aid for the beginner. It provides experience in cutting and fitting with an inexpensive fabric. It also permits the upholsterer to see exactly what the covered unit will look like. Figure 5-7 shows the inside arm of a recliner rocker covered with muslin. The ripples seen at the top indicate insufficient front-to-back stretching. There is no need to redo the muslin cover in this case since that particular problem is so minor that it will disappear when the final cover is installed.

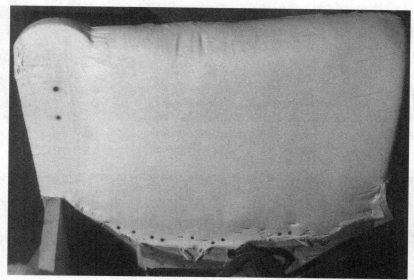

5-7 Muslin cover on inside arm. (Ripples indicate insufficient stretch of fabric, front to back.)

Figure 5-8 shows the outside portion of the arm pictured in FIG. 5-7. A section of cotton has been placed over the muslin to fill a slight void in the padding beneath the muslin cover—another benefit of using this interim cover. A light spraying of the arm section only with foam and fabric adhesive will hold the cotton in place. Figure 5-9 shows a properly padded arm and how fully the contours of the padded portions are revealed by the muslin cover.

Advantages and disadvantages of the muslin cover

Muslin covers provide many advantages:

- Provides an inexpensive experience in cutting and fitting (the muslin fabric being much less expensive than all but donated cover fabrics).
- Holds stuffing and padding in place for application of the cover, eliminating some of the "bunching" and "lifting" some novices encounter with the outer layer of cotton when moving the cover fabric over it.
- Reveals the exact contour of the padded furniture and provides for last-minute adjustments without requiring removal of a cover panel.

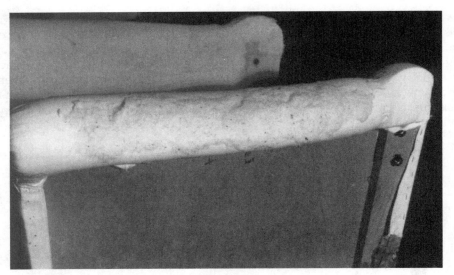

5-8 Padding added to outside of muslin cover to fill slight void. (A small amount of foam adhesive holds cotton in place.)

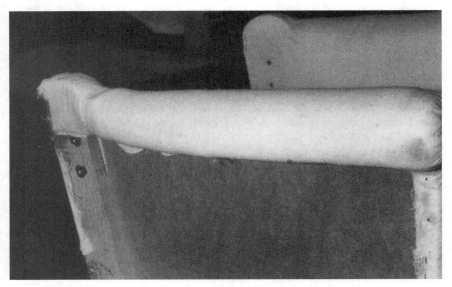

5-9 Smooth contours disclose proper padding beneath muslin cover.

The disadvantages are:

- Consumes significant time.
- Adds to material and labor costs.

6
Frames

This chapter deals with the preparations for reupholstering a unit, from a solid frame to the finished foundation. Frame design and repairs; spring installation, reinforcement, and tying; and the properties and installation of the more popular types of stuffing and padding are all briefly covered in this chapter.

FRAME DESIGN

The frame for upholstered furniture is considerably different from that of furniture with no upholstering. The following discussion will illustrate the basic principles. Only those frame members that are not self-explanatory as to their function, placement, and styling are discussed. Notice that the photos of love seat styling were taken from several different units; a group of them were being restyled at the same time.

Back construction

Back extension (FIGS. 6-1Z and 6-2C) The front contour of the back extensions should closely match the intended contour of the padded back. The forward edge of the back extension should end a little short of the the padded inside-back surface. Compare FIGS. 6-1Z and 6-2C to see how the back extension changes with a change in back styling. From this basic beginning, the upholsterer is free to determine whether the ends are to be of the wraparound styling or finished off with padded panels. In FIG. 6-1, the arm padding and cover are intended to extend beneath the back extension. The gap of about ¾ inch between the lower edge of the back extension and the arm rail gives the appearance of the arm extending into the back, which it really does. Chapter 15 discusses how the arm is upholstered using this style back.

The design in FIG. 6-2 provides for the inner arm to be finished off against, or butting into, the back upright. In this style, the back extension rests against (and may be stapled to) the arm rail. Often, the back extension has a reinforcing strip attached to the inside for increased stability (FIG. 6-1M).

Top back rail (FIGS. 6-1Y, 6-2A and Y) This design establishes the front contour of the back. A straight, flat back results from a piece of furniture having a top back rail like the one featured in FIG. 6-1. Notice that the top rail does not go all the way to the front of the extension. It is recessed enough to give room for the coil-spring suspension.

6-1 Original love-seat frame: (Z) back extension, (Y) top back rail; (X) inside back upright; (W) bottom back rail; (V) outside arm rail; (U) bottom arm rail; (T) outside arm rail (extension); (S) outside arm tack rail; (R) seat brace; (Q) wedge upright; (P) back seat rail; (O) front seat rail; (N) side seat rail; (M) reinforcing strip.

6-2 Restyled loveseat frame: (A) top back rail (crowned); (B) outside back upright; (C) back extension; (D) inside back upright; (E) arm spacing strip; (F) Lawson or rounded arm extension; (N) side seat rail; (O) front seat rail; (P) back seat rail; (S) outside arm tack rail; (U) bottom arm rail; (V) outside arm rail.

The same love seat was restyled from the flat back to a crowned back in FIG. 6-2. The coil-spring unit (which had broken components) has been replaced with sinuous springs. Notice that the sinuous springs all terminate on a straight plane at the original top rail (FIG. 6-2Y) rather than following the curved contour of the crown piece (FIG. 6-2A). This is to give the back the same tension all the way across. If the center springs are to be longer than the side ones, the center of back would be softer than the ends. If the flat back design is desired and just the coil springs replaced with sinuous springs, the top rail would have been extended to the front of the back extension, or the extensions would have been cut down, similar to those shown in FIG. 6-2C.

Back bottom rail (FIGS. 6-1 and 6-3W) First, in the case of the love seat, the front-to-back location determines the incline for the inside back. The farther forward it is placed on the *bottom arm rails* (FIG. 6-1U), the greater the incline of the inside back—the more a seated person will lean back. Second, this frame member must have its vertical placement so that padding and cover fabrics can be tucked between the bottom of it and the top of the seat spring assembly. This space is essential to permit the deck materials to be pulled beneath the bottom back rail for stapling to the top of the *back seat rail* (FIG. 6-1P). Then the inside back padding and cover are also tucked beneath the bottom back rail and stapled to the top of the *back seat rail,* on top of the decking.

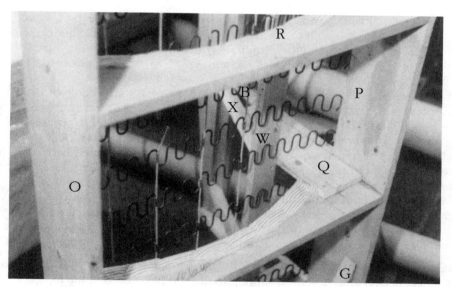

6-3 Bottom view of love-seat frame: (X) inside back upright; (W) bottom back rail; (R) seat brace, with foam and fabric padding; (Q) back wedge upright; (P) back seat rail; (O) front seat rail; (G) corner brace; (B) outside back upright.

Inside back upright (FIGS. 6-1, 6-3, 6-4X, 6-2D, and X) This element provides foundation support for coil-spring units and stability for the back bottom rail. If sinuous springs are used, the front surface of these uprights must permit normal deflection of the sinuous spring suspension without touching them. If the "straight

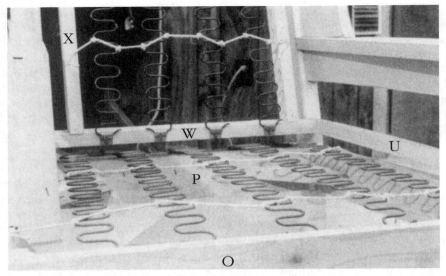

6-4 Front view of wingback chair: (X) inside back upright; (W) bottom back rail, (U) bottom arm rail; (P) back seat rail; (O) front seat rail.

stick" styling in FIG. 6-2D and FIG. 6-2X does not permit that deflection, a wedge-style upright (FIG. 6-3Q) with the front side cut concave, similar to the inside seat rails (FIGS. 6-1R and 6-3R), may be required. For some extra-wide units like 7-foot couches, this wedge-style upright is recommended to provide extra strength to the top back rail.

In the wingback chair, inside back uprights (FIG. 6-4X) are required to give a base around which to pull the inside back padding and cover. These must be spaced about ¾ inch inside the main frame members. Materials for the inside portion of the wing are also pulled between the upright and the outside frame member.

Remember that whenever two or more cover panels are pulled between two frame members, both panels should be stapled to the same member. This is to prevent "nonretrievable collection pockets."

Any upholstered furniture that has a padded side in addition to a padded back will require an upright near the outside frame members. On wider units like love seats and couches, the uprights closest to the arms (FIG. 6-2D) provide the side base for the inside back as well as spring and bottom inside back-rail support.

Outside back upright (FIGS. 6-2 and 6-3B) This element is used on wide units (love seats and couches) to give vertical support to the center portions of the top back rail and provide a base for the outside back panel when it is installed. This addition is especially critical for furniture that will be used by young people.

Seat construction

Front & back seat rails (FIGS. 6-1, 6-2, 6-3, 6-4O, and P) These rails are necessary for the front and back seat frame. Without them, the unit would be incomplete. But the way they are attached is strategic.

If the seat has front-to-back sinuous-spring suspension, and most units do, the front and back rails must be attached outside the side seat rails. Figure 6-5 shows this construction clearly. Assembled this way, the side seat rails (N) take the compressive pressures created by the pull of the sinuous springs on the front rail (O). This provides greatest strength in the direction of the greatest stress.

Now, if the seat has side-to-side sinuous-spring suspension (seldom found and only in chairs), then the side seat rails should be outside the front and back seat rails. The front and back seat rails would then bear the compressive stress, again providing the greatest strength in the direction of the greatest stress.

6-5 Top view, seat and arm assembly, wingback chair: (F) arm extension (Lawson or rounded); (G) corner brace (seat); (L) drop-spring support rail; (N) side seat rail; (O) front seat rail.

Drop-spring support rail (FIG. 6-5L) This element provides a "drop attachment" for the front end of sinuous springs. Look closely at the front rails in FIGS. 6-1 and 6-2. Notice how the springs seem to be connected to the inside of the front rail. Although the photos do not show it, a 1-x-2-inch or a 1-x-3-inch strip of wood is attached solidly to the inside of the front seat rail about ¾ inch down from the top. The sinuous spring clips are fastened to this strip, lowering the front edge of the springs. This lowering of the springs facilitates the creation of the *cushion retaining groove* (discussed in chapter 14). Figure 6-6 features a love seat with this drop-spring support.

There is no need for *back* seat rail spring supports. The sinuous springs are attached directly to the top of the back seat rail (FIG. 6-3P). This is clearly illustrated in the rear view of the wingback chair, FIG. 6-7.

Seat brace (FIGS. 6-1 & 6-3R) This brace is necessary in wide units such as love seats and couches to prevent deflection of the front and back seat rails. This is especially true for seats having sinuous-spring construction. Stress created by a series of sinuous springs attached to the front and back seat rails over a wide span is sufficient to bow a 2-x-12-inch rail (much heavier than normally found in upholstered furniture).

6-6 Love seat, showing how a drop-spring support rail facilitates the cushion retaining groove.

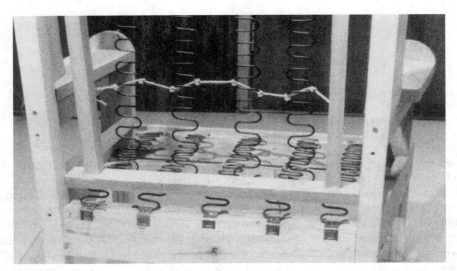

6-7 Rear view of wing back chair, showing attachment of springs to top of back seat rail. No drop-spring support is used on the backs of units.

Installing seat braces permits 1-inch lumber to be used and gives remarkable stability to the seat. With lighter-weight lumber, both cost and weight are reduced with no sacrifice in strength. The upper surface of the seat braces should be cut concave so that the sinuous springs never bottom out when a person sits down. The padding attached to the braces is to prevent tearing of the burlap undercover (thus prolonging the life of the unit).

Seat corner brace (FIGS. 6-3, 6-4, and 6-5G) This brace is a definite requirement. The addition of this simple triangular piece adds significant strength against twisting and flexing. Virtually all quality upholstered furniture is reinforced with corner braces. It is recommended that these be glued and screwed in place (not nailed).

Arm construction

Bottom arm rail (FIGS. 6-1, 6-2, and 6-4U) This rail is absolutely necessary and performs the same function as the *bottom inside back rail*—a base under which to tuck the deck padding and cover and around which to tuck the inside arm padding and cover panel. This rail is located approximately ¾ inch above the spring-suspension level. This element of the frame also provides for the attachment of any inside arm bases (cardboard, burlap, webbing), discussed later in this chapter.

Side seat rail (FIGS. 6-1 and 6-2N) Although this is really a seat component, it also receives the side attachment of the decking *and* the inside arm panels. Occasionally, some beginners have thought that the inside arm should be attached to the bottom arm rail. This is not so. That would create a horrendous garbage trap. Everything that might be dropped in the chair would find its way between the inside arm panel and the decking and end up between the frame and the outside arm panel. And to make matters worse, a person cannot reach down that far to retrieve dropped pencils, crayons, money, etc. But if the inside arm and decking are both attached to the top of the side seat rail, nothing can go beyond that point and usually can be retrieved. This same reasoning holds true for attaching the inside back to the top of the rear seat rail.

Top arm rail This rail is the base for the arm itself. From this base all arm padding and much of the contouring is developed. For example, a flat, square arm, a wrapped style, or a rounded arm can be created from the standard top arm rail. To this element may be added several other components to make arm styles easier.

Outside arm rail (FIG. 6-1V) This rail is added to the top arm rail to create a side projection to the arm, making the top of the arm wider than the stump.

Outside arm tack rail (FIGS. 6-1 and 6-2S) This rail is a necessary addition, usually located just beneath and a little inside the outer edge of the top arm rail, to give a solid base for attaching the outside portion of the inner-arm cover and the top of the outer-arm cover panel. Figure 16-2 shows an outer-arm panel being stapled to this strip.

Arm spacing strips (FIG. 6-2E) These pieces provide spacing members for rounded arms, as in the Lawson. These are usually about 1 inch square and are located in the center of the top and side arm rails. Frequently the spaces between these spacing strips and the rails are filled with foam or cotton before the basic arm padding is installed. These strips are used where rather firm arms are desired.

The arms of the wingback chair (FIGS. 6-4, 6-5, 6-7, and 6-8) have no spacing strips. An arm in this style must have the entire space between the arm rail and the upper portion of the rounded arm post filled with high-IFD foam or other firm padding. An arm of this style can be tailored to meet almost any firmness. Refer to chapter 15 for additional information on arm padding.

Figures 6-8 and 6-1 show the frame structures for a chair and a couch (actually, a love seat, but the construction for a couch is identical). Careful review of these two figures will provide about all the information you need regarding the necessary elements for quality upholstered framing.

6-8 Front view of wingback chair frame.

FRAME CONDITIONING

Quality in upholstering is reflected from the frame through the padding to the final cover. There is no such thing as an acceptable covering-up of a mistake or problem area. Irregularities will be broadcast quite reliably to the outer appearance and feel. For example, uneven spring tension will be felt if not seen; a weak suspension section will be noticed by the sitting-in feeling rather than the expected sitting on; lumps or depressions in the stuffing will eventually be felt and seen as the unit wears in. Loose joints in the frame will certainly be felt and sometimes heard. Don't take shortcuts! Do top-quality work from the frame outward.

Rounding frame edges & corners

One feature seldom found in furniture, except in those units done by quality-minded craftsmen, is rounded sharp corners and edges. Production schedules seem to be more demanding than this touch of craftsmanship. Can the difference be noticed between units having sharp corners and those that have been rounded? Eventually, in three ways: (1) wear of the padding and the cover; (2) ease of fitting, and (3) the final feel of most units. Is the extra time to round the corners and edges worth the bother? For the quality-conscious do-it-yourselfer and any true craftsman, definitely.

Figure 6-9 shows a new frame being rounded with a wood rasp. Only the springs and burlap have been installed. Old frames with sharp edges that are stripped to the frame can be rounded in the same manner as illustrated. If stripping would not normally be taken all the way to the frame, it would be unreasonable in most cases to go to the extra work of stripping to the frame just to round off the edges.

6-9 Rounding edges on a new frame with wood rasp.

FRAME REPAIRS

Broken, gouged, and weak frame components should be reinforced or replaced. The two center slats of the couch frame in FIG. 6-10 have been replaced and the two end rails reglued. Redoing the whole bottom on this unit was the best way to ensure a solid frame. Notice the use of the four bar clamps to hold the joints tight while the glue sets. Failure to clamp glued areas will result in weak, brittle joints. For best results, leave the unit under clamping pressure for several hours before working with it. Thirty minutes is a bare minimum.

6-10 Bar clamps holding repaired box-frame couch while glue is drying.

7
Buttons & channeling

This chapter deals with the many ways to finish furniture with buttons and or channels. The procedures for making the covered button, tying the upholsterer's knot to secure the buttons, and making shallow channels using 1-inch foam are all covered.

COVERED BUTTONS

Covered buttons are used extensively in upholstering. Several features contribute to their popularity. Buttons: add character and style to the furniture, are extremely easy to make, have a multiplicity of applications, and are inexpensive to make and are easy to install.

Procedure

Follow the following procedures for making covered buttons.

1. Insert button base into button-base retaining die (FIG. 7-1).
2. Place loaded button-base retaining die into the machine (FIG. 7-2).
3. Cut circles from the fabric to be used for the buttons with the cutting die; place button cover on top of the cap retaining die, finish side down; place button cap (convex side down) on top of fabric (FIG. 7-3).
4. While holding the cap retaining die suspended so the button section is free to drop down, force the fabric and button cap down into the die with the wooden plunger (FIG. 7-4). With heavier fabrics, this takes considerable pressure, so be prepared for a good push.
5. Place loaded cap retaining die on top of the base retaining die in the machine (FIG. 7-5).
6. Pull the handle down until it seats firmly. Raise the handle slightly and snap it down again with a quick motion to clinch the cap to the base securely.
7. Remove the two retaining dies from the machine and separate them. The covered button will usually drop out. If it should stick in the cap retaining die as shown in FIG. 7-6, squeeze the top and bottom of the die between the fingers, and the button will pop out. The now-empty dies, ready for reloading and the finished button (shown top up) are pictured in FIG. 7-7.

7-1 Placing button eye into button-back retaining die. The back could also be a double-prong or nail style.

7-2 Button-back retaining die in place in base of button machine.

7-3 Fabric disk (face-side down) and button cap in place, ready to be pushed into button-cap retaining die.

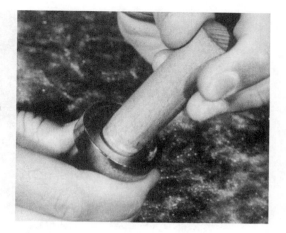

7-4 Fabric and button cap in place in button-cap retaining die.

7-5 Placing button-cap retaining die on top of button-base retaining die.

7-6 View of button-base retaining die (left) and covered button still in button-cap retaining die (right) after processing.

7-7 Top view of covered button and empty retaining dies.

Contouring

Covered buttons are often used to add contour and styling to backs and inside arms of upholstered furniture. A worker is measuring and marking the locations for buttons on a two-cushion sofa in FIG. 7-8. One of the most popular positions for back buttons is very close to the level of the arm rest, as can be seen by the position of the steel rule. Another way to determine back button position, assuming that they are installed in-line and not in a diamond or other pattern, is to measure up from the

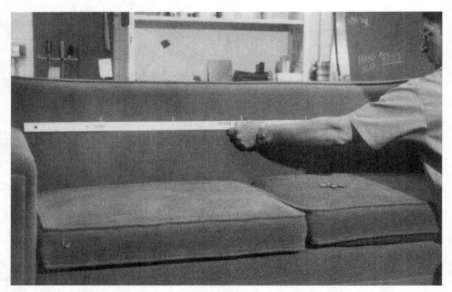

7-8 Locating and marking sofa for button placement using top-of-arm method. Notice location of bottom edge of steel rule.

cushions or down from the top of the back. Figure 7-9 shows the craftsmen locating the buttons 9 inches from the top of the seat cushions.

Buttons can also be installed on seats and seat cushions. I dislike this practice, for several reasons: (1) buttons are not comfortable to sit on; (2) in a seat, buttons seem to be crumb and dust collectors ; and (3) heavy wear pops the tops off the buttons rather quickly. For these same reasons, buttons are not often used on the tops of arm rails.

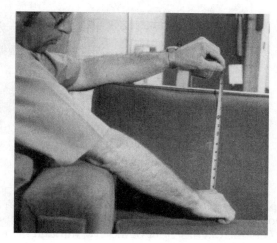

7-9 Locating and marking sofa for button placement, using vertical measuring method. This measurement could be taken up from the finished cushions or down from the top back.

Replacement

Wherever buttons are used, a prime consideration is to make it easy for the button to be secured, adjusted, and even replaced. (Occasionally a cap will pop off, leaving the less than attractive button back to decorate the unit.) To make this adjusting and replacing work possible, the following method is recommended: make a sliding loop around the eye as illustrated in FIGS. 7-10 and 7-11. *Caution:* Do not let the loop move around the twine (FIG. 7-11, left). Looping around the twine locks the twine so that no adjustment is possible and greatly increases the difficulty of replacement. Make sure the loop stays around the base of the eye as shown on the right side of (FIG. 7-11).

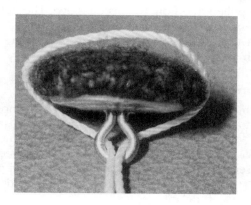

7-10 Looping tufting twine around covered button to facilitate easy replacement should that be necessary.

7-11 Left: twine will not slide through eye (undesirable). Right: proper orientation of twine loop.

This means of attaching the button makes it possible to easily replace a popped cap. All that is required is to make the new button and depress the padding and suspension while pulling outward on the old button base. Then slip the loop from around the old base, out of the eye, through the eye of the new button and around the cap. Releasing the depressed padding and suspension will usually pop the new button back into place.

Used for Cushion retaining groove

Another frequent use of the covered button is to emphasize the cushion retaining groove. This practice is becoming more popular, replacing sewing the cushion retaining groove to the springs and burlap base. Figure 7-12 shows starting a button, using a tufting needle. All strands of the tufting twine are held between the thumb and fingers until they begin to penetrate the fabric; then the worker releases the twine and works the needle with both hands to force it through all padding and stuffing until the needle has gone the full length or it can be felt that the twine has been pulled free from the needle's eye. The button will then be tied to the desired depression of the seat and deck joint.

A third use of covered buttons is to create depressions in the tufting process, which is illustrated in chapter 8.

Securing covered buttons

One method for securing covered buttons is shown in FIG. 7-13. This method is used when more depression is desired than can be achieved by the procedure shown in FIGS. 7-14–7-26. In this application, the loop formed after the upholsterer's knot is tied is hooked around a number-12 tack that has been driven in most of the way (FIG. 7-13). Tension is adjusted by pulling on the "slipping" strand of the twine. Notice in FIG. 7-13 that the operator is holding the outside back panel and foam padding out of the way for clarity.

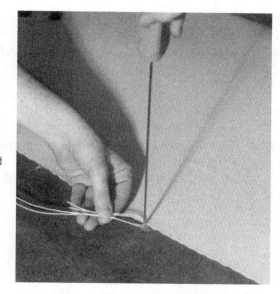

7-12 Inserting tufting twine at desired location along a cushion retaining groove on a couch that received a new decking.

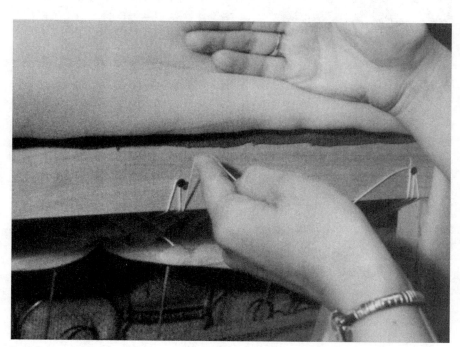

7-13 Attaching tufting twine loop formed with upholsterer's knot to tack or nail in frame to anchor covered button. (Operator is preparing to make a double wrap with long end of twine to prevent cutting main loop when tack is driven in.)

Tying the upholsterer's knot

The most practical and versatile way to secure buttons is with the *upholsterer's knot*. This knot is easy to tie, forms a one-way slip knot that is ideal for adjusting tension on the ties, and is secured permanently with a simple overhand knot once the desired tension is achieved.

After all buttons to be tied have been installed (refer to FIG. 7-12), follow the sequence below for an almost guaranteed success in tying them off and getting even tension.

1. Grasp a strand with each hand and pull alternately to ensure that the twine slips freely through and around the eye of the button (FIGS. 7-14 and 7-15). Pull both strands firmly and simultaneously to ensure that the eye of the button has penetrated the cover fabric. If it appears difficult to penetrate the cover, pull the button free just enough to get to the hole easily and carefully enlarge the opening by cutting a tiny slit with a knife or shears or wedging the opening larger with a stuffing regulator. A pull on both strands of button tufting twine should now be sufficient to slip the eye of the button through the fabric.

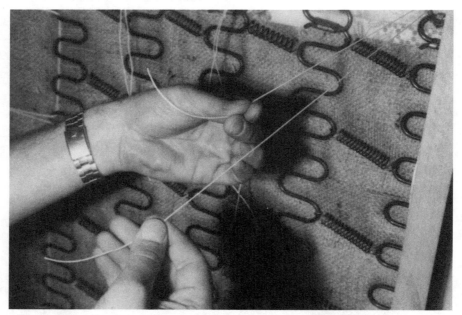

7-14 Making sure that tufting twine will slide easily through button eye.

2. Grasp the "slipping" strand (which I prefer as the longer of the two) between the third and fourth fingers and the palm of one hand (FIG. 7-16). Drape the "tying" strand over the top and around the fingers and reach under to grasp it with the right hand, as shown in FIG. 7-16.

3. Pivot the left hand toward the right, as shown in FIG. 7-17, while holding the "tying" strand (the shorter of the two) to the side. Notice the opening that

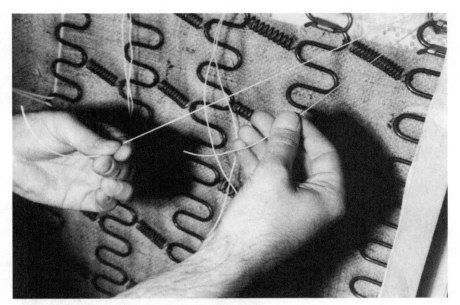

7-15 Adjusting tufting twine for tying upholsterer's knot.

7-16 Beginning of upholsterer's knot, creating loop around fingers.

has been formed by the first and second fingers of the left hand (FIG. 7-17). That is essential.

4. Thread the free end of the tying strand through the opening just in front of the first and second fingers of the left hand (FIG. 7-18).

5. Reach underneath with the right hand and pull the free end through (FIG. 7-19).

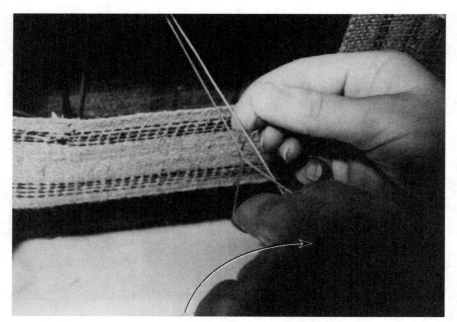

7-17 Rotate loop, rolling hand toward the hand holding short end of twine.

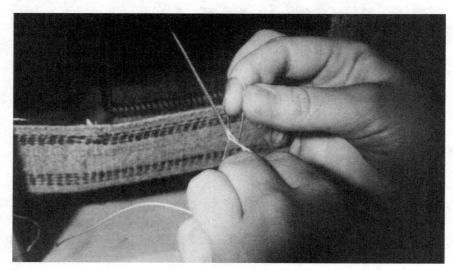

7-18 Thread short end of twine through loop.

6. Rotate the left hand back so the palm is up. The view should appear as in FIG. 7-20. If it does not, undo the knot and begin again with step 2.

7. Tighten the knot slightly by pulling on the tying strand (arrow, FIG. 7-21) and a clearly defined "figure eight" will appear, as illustrated in FIG. 7-21. The slipping strand is held taut at this point.

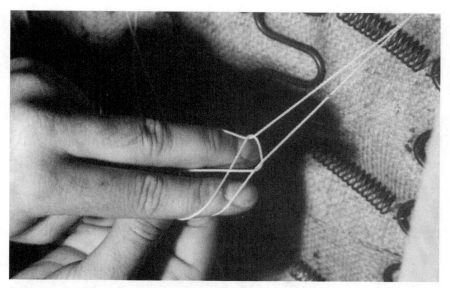

7-19 Reach under, grasp short end while holding on to long end.

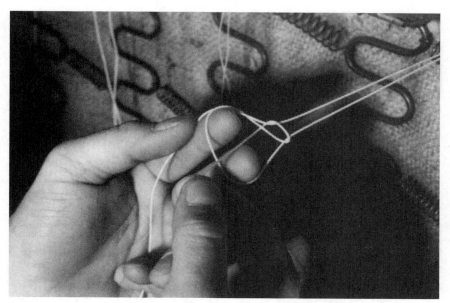

7-20 Rotate hand back, slip fingers out, and begin to tighten knot.

8. Finish tightening the knot (FIGS. 7-22 and 7-23). Do not tighten the knot so much that it makes the slipping action very difficult. Figure 7-23 depicts pulling on the slipping strand (arrow) to reduce the size of the large loop that exists between the knot and the spring assembly.

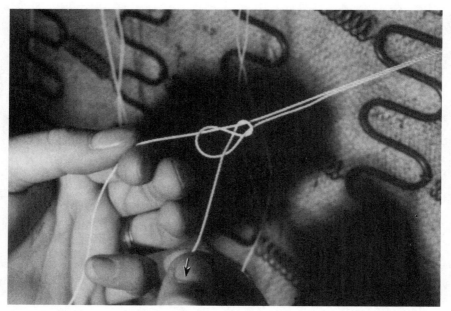

7-21 Pull on shorter end (arrow), and you should see a figure eight forming around the long or sliding strand.

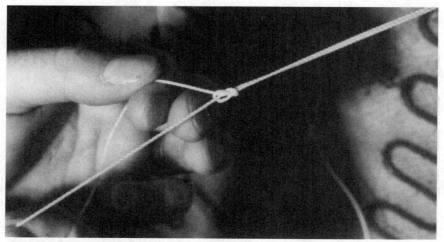

7-22 If properly done, the upholsterer's knot will appear like this as it tightens.

Securing the upholsterer's knot

One of the most popular ways to secure the upholsterer's knot is to place a wad of cotton scrap or rolled fabric between the strands (FIG. 7-24), slipping the knot to create tension (FIG. 7-25). Snug all buttons about the same and go around to the front side to see if the tension is as desired. If more depression is desired, increase tension. Because the knot is a one-way slip knot, it will retain the tension as applied. If

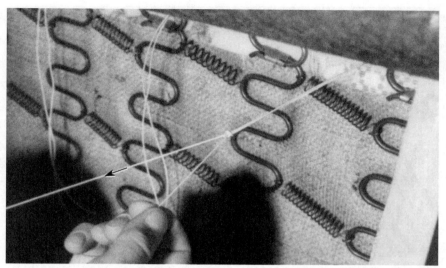

7-23 Once snug, pull on the sliding strand (arrow) to slip knot closer to foundation.

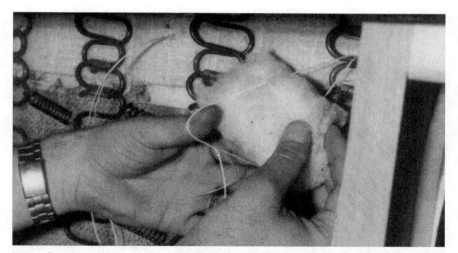

7-24 Open loop; insert scrap cotton, fabric, or Dacron to provide an anchor around which to lock the knot.

the tension is too great, hold the slipping strand firmly beneath the knot and pull backward on the knot itself. *Caution:* Pulling on the tying strand will only tighten the knot and make it more difficult to slide it backward.

Once the desired tension has been achieved for all buttons, secure the knot by (1) holding the fingers of one hand against the outside of the knot while tightening the tying strand further; then (2) make an overhand knot (the kind used to start tying shoelaces) and tighten this against the upholsterer's knot. That's what the craftsman is doing in FIG. 7-26. Look closely. It is hard to see the white twine against the whiteness of the cotton.

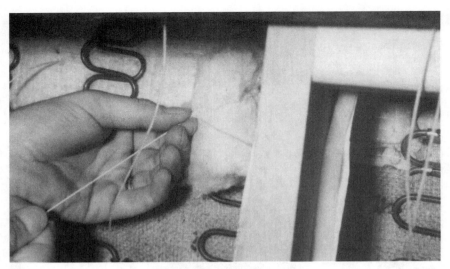

7-25 Snug knot against stuffing material, adjust all buttons evenly before locking them in place.

7-26 Lock the upholsterer's knot by tying an overhand on top of it.

The strands shown dangling vertically from the top of FIG. 7-26 are remnants of buttons previously secured the alternate way. Figure 7-13 illustrates that method of anchoring them to a frame member.

Some upholsterers have achieved such an accurate feel for the desired tension that they pull the strands at the same time and staple them to the frame. No adjustment is possible—and usually not necessary because of the skill they have developed. The latter method for securing buttons is usually identified by the strands being stapled, folded over, and stapled again to prevent any slippage as the spring action takes place.

CHANNELS

Channeling adds an air of elegance beyond buttoning alone. Figure 7-27 shows a couch with a plain buttoned back ready to be reupholstered. The same unit with a buttoned channel back is shown in FIG. 7-28.

There are differences in shapes, widths, depths, construction methods, stuffing techniques, and materials from which channels are made. At one time, it was com-

7-27 View of a buttoned plain back prior to reupholstering.

7-28 Reupholstered couch with buttoned channels, in 1-inch foam.

mon to pack stuffing materials into sleeves sewn of a cover fabric and a less expensive backing material. Stuffing materials were of hog hair; cotton (shredded or rolled); chopped foam, Dacron, excelsior (a woody fiber), and sometimes paper; or a combination of all of these materials. However, most channels are now being constructed from flat foam sheets.

Channels are most often used on the backs of upholstered furniture. Occasionally they are used on inside arms and less frequently on seats and cushions. For this reason, the discussion on channels deals with a back section.

Elements of a channel sandwich

A channel unit usually consists of three parts, bonded and sewn together (hence the term *sandwich*): (1) a lightweight, inexpensive backing fabric, (2) the foam padding, and (3) the cover fabric. One of the more popular ways to create this sandwich is by using 1-to 3-inch foam between the backing and cover.

Padding for the channel sandwich

The first essential step in creating a channel unit or sandwich is to obtain the major measurements—width (side to side) and depth (top to bottom)—of an adequately padded unit (refer to chapters 13, 14, and 15). Because a channel sandwich is itself a bit of padding, TABLE 7-1 can help you determine how much padding should be applied to a back before measurements are taken. To make this determination, touch the tips of the thumbs together, fingers spread wide, and press the hands onto the padded back section, then use the described conditions in TABLE 7-1 to determine if sufficient padding has been applied.

Table 7-1 Minimum padding prior to measuring for channel components.

Foam thickness	Minimum padding acceptable
½"	No discernment of foundation elements (springs etc.).
1"	Springs barely distinguishable, no discernible lumps or depressions.
1 to 2"	Springs may be felt but no lumps or depressions should be felt.
2"	No "pronounced" depressions or lumps is all that is required.

(Pad the "base" until the above conditions exist.)

Determining the size of channel elements

1. Determine the size for the sheet of foam of the desired thickness and density for your application. Normally it will be a bit *oversized*, the amount depending upon the styling desired and the thickness of the foam to be used.

   ~Record the greatest straight-line measurements of the width and depth of the area to be channeled. These are the crown-to-crown dimensions, (the distance from one side to the other, measured around the curvature of the padding, not the straight-line distance).

~Take a scrap or section of the foam that is to be used in the channel and using a piece of fabric on top of it, form the foam around corners or edges to where it will be terminated. Make adjustments until it forms and ends where you want it to. Hold the flat portion of foam in place against the back and let the section that forms the rounding return to its basic flat position. Measure from the outer edge of the foam to a point in line with the crown-to-crown measurement taken in the preceding first step. Add double these measurements (one for each end) to the crown measurements taken earlier, and the length (side-to-side) is established.

~The width (top-to-bottom) can be determined more easily by measuring with a flexible tape from the tacking point at the top rail down to the deck. Allow a space between the tape and the padding approximately the thickness of the foam to be used, then add an extra 3 to 4 inches to pull between the bottom back rail and the deck. The width is established.

2. Measure and cut the backing fabric 2 inches longer and wider than the foam.

3. The width (depth) of the cover panel is made 2 inches wider than the foam. To this, a *stretcher* wide enough to reach the rear seat rail will be added at the bottom. Or if cover fabric is to be used instead of sewing on the stretcher, add enough extra cover to reach the rear seat rail. This choice is a matter of economy.

When making the fabric layout, if sufficient cover fabric is available so it doesn't create extra expense in cutting other panels, use it for the stretcher portion. This avoids the additional, though minor, sewing of the stretcher. The latter choice is often preferred when doing single units where yardage has been ordered according to the standard charts. To calculate the cover length, use this basic formula:

$L = 2 + F + 1.6dn$, where
L = total length (side-to-side) of cover
F = length of the foam
d = depth of each channel cut (into the foam)
n = number of channel cuts

Creating the channel sandwich

Now that the component sizes have been determined, proceed with the construction of the channel sandwich.

1. Lay the foam on a flat, clean work surface. It is a good idea to have some scrap paper underneath the foam, extending beyond the edges, to protect the work surface from adhesive overspray.

2. Smooth the backing fabric over the foam so that it overlaps evenly on all sides, similar to that shown in FIG. 7-30.

3. Carefully fold about half of the backing over and lay it on the top of itself, keeping everything relatively smooth and free of wrinkles (FIG. 7-29).

4. Spray the exposed underside of the backing and the exposed portion of the foam with fabric and foam adhesive. With the backing folded over itself, as

7-29 Spraying foam and backing fabric to bond the two together.

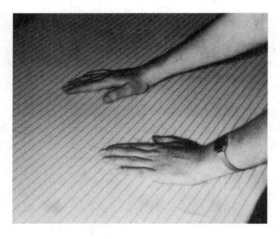

7-30 Smooth out all wrinkles while bonding backing to foam.

illustrated in FIG. 7-29, the full half of the backing and half of the foam can be sprayed on one step. Let the wet look disappear from the adhesive.

5. Adhere the backing to foam. Carefully suspend the sprayed half of the backing over the sprayed half of the foam. This is best done by grasping one corner of the backing in each hand and pulling it straight out over the foam without letting any portion of the sprayed half touch until it is all straight. (There is usually enough frictional resistance of the unsprayed half to keep the fabric from moving.) Then let the backing slowly down to contact the foam, beginning at the center and progressing outward. If done carefully, the backing can be smoothed against the foam with very small or no wrinkles. That is what is shown in FIG. 7-30. A light rubbing action bonds the two together. Any attempt at this point to move the fabric with respect to the foam will probably result in tearing the foam, so don't try it.

6. Fold over the second half of the backing and repeat steps 4 and 5.

7. Lay out and mark the foam for the channels. Turn the backed foam over so the foam is up. Determine the width, number, and location of channels. The width and number of the channels is a matter of choice; both depend somewhat on personal preference and the width of the unit. Those pictured in FIG. 7-31 are 9 inches wide. Centering is also a matter of choice. At times, one of the channel *seams* will be center. The other alternative is to center the channel itself along the centerline of the unit. Mark the foam with a felt-tip marker.

8. Cut the channels. A steel rule and a sharp razor-blade cutter are the best tools for this operation. The first cut is made using the steel rule as a straightedge, cutting the lines marked (FIG. 7-31). Make no attempt to cut deeply with the first pass. The second cut is made by slightly opening the first cut, not using the straightedge. This is done by pressing and pulling the top of the foam to one side slightly, as illustrated in FIG. 7-32. Cut the foam to the depth desired (½ to ¾ inch is the average depth in 1 inch foam). *Warning:* Do not cut all the way through the foam and into the backing.

7-31 Use steel rule to guide first cut in foam.

9. Form the cover to the channels. Start this process with the center channel and move alternately outward to the ends. Fold the cover in half, face-side to face-side, and locate the fold line, overlapping the edge of the center channel by half the depth of the cut, as illustrated in FIG. 7-33.

10. Open channel cut and press folded cover into it. (FIGS. 7-34 and 7-35). When the folded fabric is pressed smoothly into the cut, let the foam come back into place, as seen in the upper portion of FIG. 7-35.

11. Smooth fabric into channel. After the entire length has been put into the fold, open the cover fold and smooth the fabric into the groove with the sides of the fingers (FIG. 7-36).The finished forming should look like FIG. 7-37.

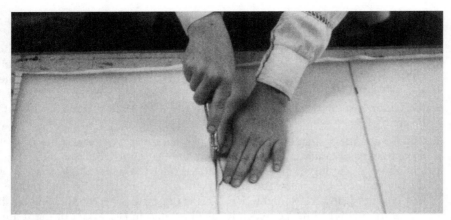

7-32 Open as you slide to desired depth. (Multiple shallow cuts are much better than trying to achieve full depth in one cut.)

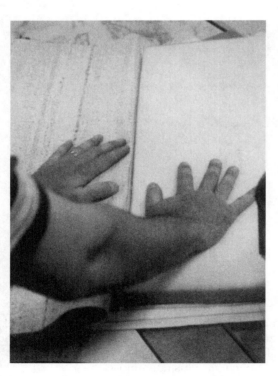

7-33 Fold cover fabric to slightly overlap channel cut. Work from center channel outward to both ends.

12. Pin channel crease in place before moving to the next one. Insert a skewer along the centerline of the crease so that it will hold the cover securely to the backing fabric (FIG. 7-38). Lifting the channel sandwich from the bottom to create a slight convex curvature along the seam line will make it much easier to get the skewer to catch the backing. Place the outside pines 2½ to 3 inches from each edge. Three or four pins will be required for each crease.

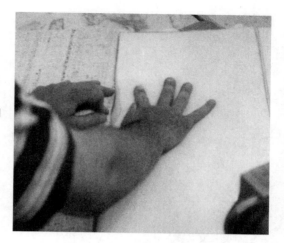

7-34 Open cut and press folded cover into it.

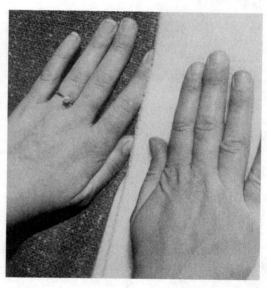

7-35 Let foam relax against tucked-in fabric to hold in place.

13. Sew the channels. Beginning with the center crease, sew the sandwich together along the centerline. Taking time to make sure that the needle of the sewing machine is exactly on the centerline as the channel is sewn will pay big dividends. It is the only way to have a smooth channel. Just after starting the stitch, pull out the end pin. Sew along the crease to within a couple of inches of the next pin, then remove it. *Warning:* Sewing over the skewers could be dangerous! Broken needles sometimes fly with blinding speed.

The crease being sewn in FIG. 7-39 is the last one on one end; otherwise there would a large roll of material being forced through the throat of the machine to the right of the needle. Opening the fabric slightly, as depicted in FIG. 7-39, makes it easier to keep the needle and the walking foot from catching the sides of the cover material, thus aiding in maintaining a straight seam.

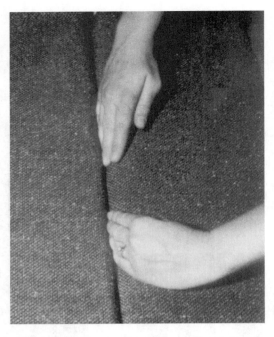

7-36 Lay cover open and smooth fabric deeper into slit. Do this for each channel before continuing to next one.

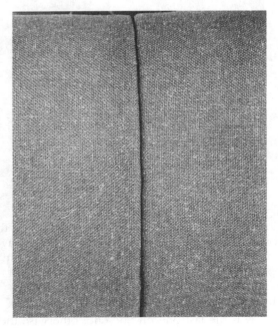

7-37 A properly prepared channel in 1-inch foam prior to pinning and sewing.

7-38 Pin each channel fold before tucking in the next one.

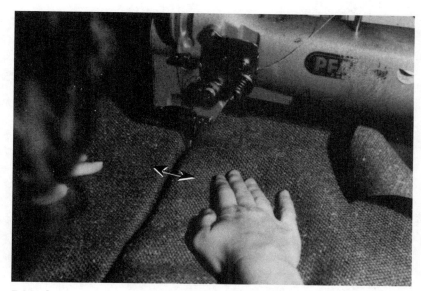

7-39 Sewing the channel seam. Slightly opening seam lets you see better to keep the stitch going along the center of the stitch.

8
Tufting

Tufting is a means of adding significant beauty to the backs of chairs, love seats, couches, headboards and footboards for beds, and some bench seats. *Tufting* is the practice of pulling, folding, or sewing in a contoured pattern, usually three basic geometric shapes—diamond, square (biscuit), or triangle. The latter two practices, folding and sewing, are explained and illustrated later. Pulled tufting is so basic that no detailed information is necessary.

All tufting procedures involve pulling the apexes of the desired pattern into the padding. One procedure, the most popular, is to use covered buttons to create the pattern depressions. Buttons are covered with a fabric, most often the same as that of the cover, then applied from the front or top side of the unit and pulled into the padding to the desired depth. These buttons are either the eye type, affixed with twine, or the prong type, affixed with two metal strips spread apart to hold the button in. Pronged buttons are generally used where the tufting is mounted onto a solid base like wood or metal.

The second procedure is to attach a twine to the underside of the fabric where the lines of the pattern intersect. The twine is invisible from the top. This procedure is excellent for eliminating the hard-button feel and is used primarily in restaurant, vinyl-covered bench seating. This "bottom-pulled" procedure usually involves sewing in the pattern.

True tufting requires considerable extra planning and work. For this reason, many upholsterers charge an extra $3 to $5 per tuft. As the two tufting methods (sewing and folding) are explained, it will become evident why this extra charge is applied.

FACTORS TO CONSIDER IN TUFTING

For practical reasons, I prefer not to tuft (or channel for that matter) seats. Tufting creates crumb, dust, and spill collectors, requiring constant cleanout. Seats receive much more wear than backs and buttons often pop off. Buttons, which can be felt, depending on the depth (or lack thereof) of the tufting, along with the bumps and valleys don't make a very comfortable seat.

When working with vinyls, a couple of other considerations should be noted. The perforation effect (discussed in chapter 9) causes seam separation. Folds and creases result in cracking earlier than normally expected (vinyls do not have a long flex-fatigue life). Seat tufting with vinyls, then, results in shortened upholstery life.

If you recognize these considerations and you still want tufting, go for it. The practice is popular. It looks very nice, and some period furniture requires tufting.

TUFTING DESIGNS

Three basic designs dictate tufting. The two most popular tufting designs are discussed. The third tufting method, *triangular*, is generally a pulled procedure and not a true tuft. Tufts forming diamond patterns is called *diamond tufting*. Tufting creating a square pattern is known as *biscuit tufting*.

TUFTING METHODS

Modern filler materials used in the tufting process come in two forms: *sheet* urethane foam and *chopped* urethane foam. For that reason, we will discuss methods to create tufts using foam. (In earlier days of upholstering, prior to the 1980s, some of the stuffing used for tufting and channeling was of loose fibers—horse and pig hair, various plant fibers, and some of the mosses; these are no longer used.) Four basic tufting methods will be presented.

Top-stitched In this less-used method the desired pattern (diamond or biscuit) is marked and cut into a foam, usually no more than 1 inch thick). The foam is backed with an inexpensive fabric, using the same principle as the top-stitched channeling described in chapter 7. The cover is placed on top, and the sandwich is sewn together. The sewn pattern is then buttoned at each of the intersections to a chair or couch on which the majority of padding is already in place. The tufted sandwich may be thought of as the final layer. Calculating the sizes of the backing, foam filler, and cover fabric is exactly like that for channeling. That is, 1.6 times the number of tufts in each direction (across and up-and-down), times the depth of the tufts, plus 1-inch minimum per side for pulling the cover into place for stapling to the frame.

Sewn tufting A second method is to sew the pattern into the cover so that when buttons are installed, a natural pocket or recess is formed to permit the buttons to be depressed significantly into the padding. This style is especially suited for pillow backs and arms stuffed with chopped materials, foam, foam and cotton, down, or other loose fillers. This style will be explained in more detail below.

Folded tufting The third method, the real art of the craft, is to mark the desired pattern on *unbacked* foam, drill holes at the intersections partially or completely through, slice partially down the tufting lines, and button and fold the cover fabric into each junction. This method will be explained in some detail below.

Pulled tufting This is the simplest of the tufting processes. No folding or sewing to achieve the pattern occurs; it is all pulled into place. An oversized cover fabric is placed on top of a 3- or 4-inch foam. Buttons are then *pulled* into the sandwich to create the desired pattern. The cover panel is usually cut oversize by 1 inch per tuft. For example, if there are three tufts across the width and four tufts up and down, the cover panel will be cut 3 inches wider and 4 inches longer than normal. This will allow the necessary material to form the tufted contour.

FOLDED TUFTING

This is the real tufting art—folding the fabric to create a smooth tufted pattern. This procedure requires that the fabric have some degree of stretchability. For that reason, velvets, some vinyls, and leathers will be tufted this way. Figure 8-1 shows a

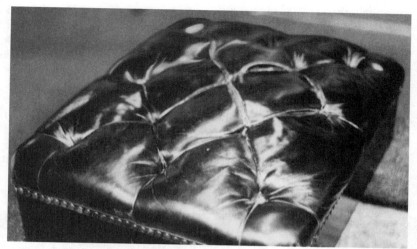

8-1 Folded diamond-tufted (leather) piano bench seat to be reupholstered. (Notice the opening of one of the tufts.)

leather-covered piano bench ready to be reupholstered. Notice that toward the rear, one of the folds is split open. This is typical wear of this type of tufting. Proceed in this manner:

1. Cut a piece of foam of the thickness desired, in this case 3 inches, so it will overlap each side by ¼ inch. (If the bench seat measured 13 × 21 inches, the foam would be cut 13½ × 21½ inches.) Mark the centerlines for the width and length, FIG. 8-2. Dimensions for the cover fabric will be detailed in (Fabric and Foam Sizes.)

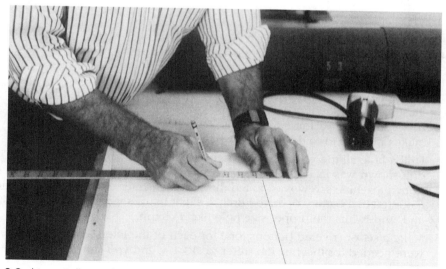

8-2 Lay out diamond pattern on foam after scribing centerlines.

2. Determine the pattern size and how the pattern is to be oriented. A popular size for diamond tufting, for both seats and backs, is 4½ inches wide by 7 inches long. Look again at FIG. 8-1. You will see that the design calls for two widths of the diamond pattern across the center. In FIG. 8-3, the worker has measured 2¼ inches from the centerline (this locates the centerline for the first diamond) and is now marking down 3½ inches. Where the pencil is will be the lower point of the first diamond. Continue marking off the dimensions of each diamond, then connect the markings using a felt-tipped marker, as illustrated in FIG. 8-3. (You will probably notice that toward the bottom one of the lines is out of place. A second worker was helping and mismarked the width of the second diamond by 1 inch. Fortunately, it was noticed before all lines were connected.)

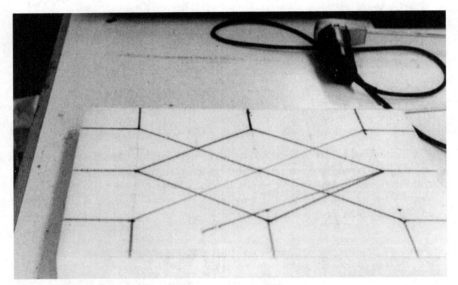

8-3 Diamond pattern laid out on foam, ready for drilling holes.

3. Mark perpendicular lines outward to the sides and ends of the foam, as illustrated in FIG. 8-4. If possible, it is best to leave 2 inches or more between the last diamond intersection and the edge of the foam. This makes it easier to fold in the corners.

4. Drill a hole all the way through the foam at each intersection (FIG. 8-4). The drill shown was made from the brass ferrule from a broken wooden leg. A ¼-inch machine screw was bolted through the center of the (now²) top and mounted in a battery-operated drill. A slight bit of sharpening of the cutting end, and it cuts the proper size hole like a charm.

5. Make a set of covered buttons, one for each of the intersections. Thirteen were needed for this first job (refer to FIG. 8-4, and you can get the count quickly).

8-4 Drilling holes through foam with a hole saw made from a brass ferrule from a broken chair leg. A ¼" machine screw has been centered in the end to serve as a mandrel.

6. Cut the perpendicular lines down, about halfway into the foam. Figure 8-5 shows one man using a regular hacksaw to do this. The cuts may also be done with a razor knife (my preference), as illustrated in the channeling process, chapter 7.

7. Begin at the center of the "top" end , insert a pronged button through the fabric at the intersection, and push it all the way into the base board. Spread the prongs out. Continue this process, alternately working from side to side until all buttons are in place. Make all angular and perpendicular side folds so they face downward, (FIG. 8-6). Vertical or end folds should be made in one of three ways: (1) so the folds all lay in one direction (the preferred method when there is an odd number of end folds); (2) so the folds face outward from the centerline (as in FIG. 8-6); or (3) so the folds face inward toward the centerline. Be consistent in how this is done. Notice in FIG. 8-6 how all the diamond folds are facing toward the camera. The end folds are facing outward from the center. Of course, a center fold, as in this case, must face one direction or the other. The art behind the folding process will be illustrated in greater detail a little later.

8. After all folding is in place, stay-tack each of the perpendicular folds to the base. Then, working from the center toward the folds of each tuft section, staple each side section in place. Remove the stay tacks as you approach the perpendicular folds, take up the excess, and restaple. Figure 8-6 shows the new piano seat with everything folded and stapled in place. Part of the end fabric has been trimmed close to the staples.

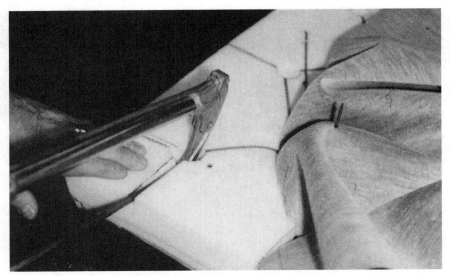

8-5 One method of cutting perpendicular slits, using a hacksaw in foam for outside folds.

8-6 End view of diamond-tufted piano bench showing that all folds are facing the same direction.

9. Cover the staples with a matching vinyl strip and install enameled or decorative tacks as shown in FIG. 8-7. Premade stripping is available commercially, or you can fold and glue your own. The tack hammer is tipped with a white nylon. This prevents marring of the decorative (in this case painted) tack heads. Notice also that stay-tacking is used to hold the vinyl strip in a straight line.

    Nylon-tipped hammers may be purchased or you can tip an old tack hammer rather easily. Obtain a piece of ½-inch-diameter nylon rod from a

8-7 Installing enameled tacks over ¾" vinyl stripping, available commercially. A nylon-tipped hammer is used.

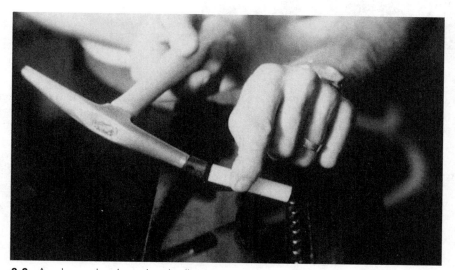

8-8 A nylon rod makes a handy alignment tool for any tacks that bend slightly during setting.

local plastic supply outlet. Cut off a ½-inch piece. Drill a hole ¼ inch deep (about ⅛-inch diameter) in the center of one end of the nylon rod and in the center of the face of the tack hammer. Get a piece of steel rod [do not use aluminum; it is too soft; a heavy coat hanger can sometimes be used]. That makes a press fit with the drilled holes. Cut this rod just a little short of ½ inch. Taper each end slightly and drive it into the hammer face first. Then drive on the nylon tip.

If you inadvertently bend a tack slightly sideways, you can sometimes tap it into proper alignment using a short section of nylon rod, the same type used as a tip for the tack hammer (FIG. 8-8). As with the nylon tip on the tack hammer, the nylon rod does not chip the paint from the decorative tack. The hammer in FIG. 8-8 has a homemade black Delrin (acetal plastic) tip very similar to nylon.

10. Continue the tacking until the job is completed. Figure 8-9 shows an end view of the progress along the last side.

8-9 Stay-tacks holding vinyl strip in alignment as tacking nears completion.

FABRIC & FOAM SIZES

As with channeling, the pattern marked on the cover must be larger than the pattern on the foam. Like many other things in upholstering, there is more than one way to do the job. TABLE 8-1 gives one approach of how much to add for each tuft. In this system, the width is increased by ¼ inch *more* than the length.

Upholsterers find easy ways to keep track of things. For example, they choose to remember only one dimension increase (for 2-inch foam, add 1½ inches for width,

Table 8-1
Diamond tufting
*Increased dimensions.

Foam thickness	Add to width	Add to length
1"	1¼"	1"
2"	1½"	1¼"
3"	1¾"	1½"
4"	2"	1¾"

*Add to each tuft for cover fabric pattern.

1¼ inches for length). That then becomes the standard. For each inch of increase or decrease in foam thickness beyond the 2-inch standard, change the fabric dimensions by ¼ inch.

For a basic diamond pattern 5½ inches wide and 7 inches long to be tufted in 2-inch foam, the width would be increased by 1½ inches and the length would be increased by 1¼ inches. This means that in the diamond pattern applied to the fabric, each diamond is 7 × 8¼ inches.

If the basic diamond pattern of 4 × 6¾ inches is chosen and is to be tufted in 3-inch foam, TABLE 8-1 shows that the width would be increased by 1¾ inches and the length increased by 1½ inches. The diamond pattern on the fabric would be 5¾ × 8¼ inches.

A second system exactly reverses the increase for length and width. It is interesting to note that the resulting tufting seems to come out the same. In other words, the length is increased ¼ inch more than the width. But the same standard is used for both systems—1½ inches and 1¼ inches for 2-inch foam. Choose the system you wish to use. Just be consistent; that is, make all diamonds the same proportion.

Figure 8-10 shows the markings for a new diamond-tufted seat. Shown in the photo are the particleboard base (A), 2-inch foam pad (B), and the underside of the vinyl cover (C). The three elements were measured and marked in this manner:

1. The foam pattern was laid out, marked, and drilled. The diamond size in this case was 4¼ inches wide by 6¾ inches long.

2. The base (A) was placed beneath the drilled foam (B) and the centers of the holes marked onto the base board with a felt-tipped marker.

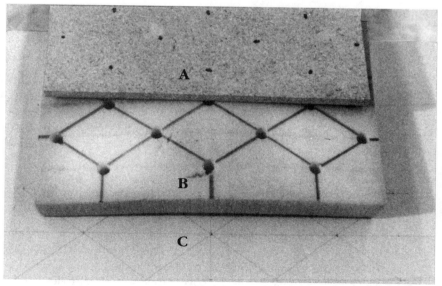

8-10 Markings for a diamond-tufted bench seat: (A) particleboard seat base; (B) 2-inch HR, 50 IFD foam; (C) back side of vinyl cover fabric. (Notice that the pattern on the fabric is significantly larger than on the foam.)

3. The fabric pattern (C) was then laid out, making the diamonds 6 inches wide and 8 inches long. The center lines of the three pieces in FIG. 8-10 have been lined up so you can see the difference in the size of the pattern on the foam and the fabric. You may also notice that the dots on the particleboard are the same size as the foam pattern. This is as it should be. A seat of the construction type in FIG. 8-10 is designed to fit inside a frame, resting on supports attached to the insides of the frame.

FOLDED TUFTING WITH SEWN PERPENDICULARS

Where a seat is to be tufted and the pattern leaves less than 2 inches at the corners to be folded in, a combination of the folding and sewing process is recommended. Take another look at FIG. 8-10. Can you see that the longer dimension of the diamonds terminates quite a bit less than 3 inches from the edge of the foam? To try to fold this short of a span and have it hold would be very difficult. Take a ¼-inch tuck on each side of the perpendicular folds and sew them in. Three sides of the completed sewing are shown in FIG. 8-11.

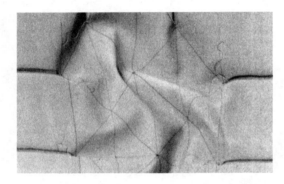

8-11 Back side of vinyl cover for folded tufting with sewn perpendiculars.

1. Push the prongs of the first button through the fabric and press it all the way through the foam and the board hole. Spread the prongs, and you will have a beginning tuft (FIG. 8-12).

2. Put in a couple more buttons, pressing each one all the way to the base, and spread the prongs out to lock the buttons in place. Figure 8-13 shows what the tufting looks like with three buttons in place.

3. Push a few more buttons through the cover at the points of the diamond pattern (FIG. 8-14), fasten them to the base, then fold the first few tufts into place. Figure 8-15 shows the new seat with a few of the tufts folded (just to the left of the worker's hands) and the "wrinkles" of another tuft being rolled out. Here is the strategy of folded tufting—rolling out wrinkles in the cover fabric without picking up any of the padding, then laying the "fold" down in the proper direction. Figure 8-16 shows the fold ready to be laid in place. As you look at FIG. 8-16, you can see how the cover fabric would be "rolled" between the thumbs and forefingers. Notice also that the wrinkles have been removed from the side the fingers are on and that the fold is being laid in that direction, matching the other folds, all with the fold laying in the same direction.

8-12 Beginning of a folded diamond tuft; this is what the first tuft looks like when the double-pronged button is installed.

8-13 Diamond tuft: first three buttons installed before any "folding" is done.

8-14 Pushing the double-pronged fastener through one of the outside points of a diamond tuft.

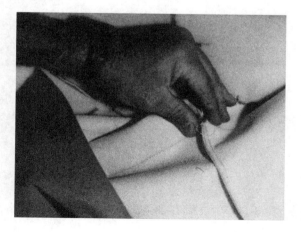

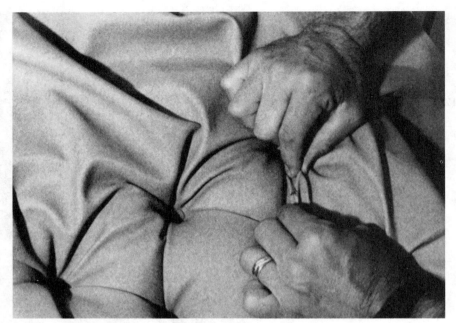

8-15 Worker lifting vinyl to clear wrinkles and roll fold into place. (Notice those that have been smoothed and folded.)

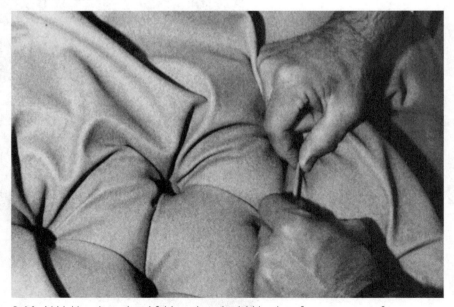

8-16 Wrinkles cleared and fold ready to be laid in place for one more tuft.

4. Continue this process until all folds have been made for the diamond pattern.

5. Arrange each side fold so it will be perpendicular to the edges of the base and stay-tack to the *bottom* of the base. (Remember, this type of seat construction will fit within a framework.) Stay-tack all but the cornermost folds, leaving them to do last. Even though these perpendicular folds have been sewn, there will probably be additional fabric to roll into them.

6. Pull down the center portions of each of these side tufts to make an even, snug tuft. These should be slightly lower in profile than the center tufts.

7. Remove the stay tacks at the folds, roll in any additional slack fabric, lay the fold in the same "down" direction as the rest of the tufting, and tack to the bottom of the base.

8. Cut off the excess fabric within ¼ inch of the staples and place the tufted seat in its frame.

SEWN TUFTING

The old vinyl cover for a recliner has been removed and laid out to show the sewn tufts in FIG. 8-17. The tucks were about ¾ inch each, creating deep tufting in chopped foam backing. Notice how high the tuft points are standing. This indicates that the tufting was rather deep, about 4 inches. Here is the procedure for creating sewn tufting:

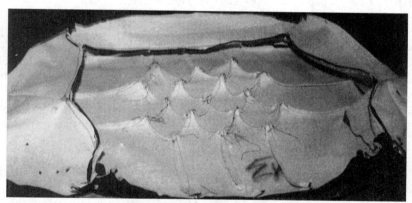

8-17 Old vinyl cover, showing fully-sewn tufting (from a recliner).

1. Lay out the desired pattern on the back side of the fabric panel. Figure 8-18 shows two workers completing the markings for a sewn diamond pattern.

~Mark the horizontal and vertical centerlines (notice the vertical line extending from top to bottom, even through the centers of the diamonds. The horizontal centerline is very difficult to see.)

~Draw in the diagonal centerlines that will create the diamond pattern.

~Scribe circles at the junctions where the buttons will be placed. A template is handy, like the quarter being used by the worker at the upper right of the photo.

~Mark the sewing lines, parallel to the diagonal centerlines, and round these to a point at the circles (shown in process at the lower left in FIG. 8-18). Virtually all these sewing lines can be sketched freehand if a bit of care is taken.

8-18 Laying out pattern for a sewn diamond tuft. A quarter is used to trace circles at tuft intersections while another worker sketches in sewing lines.

2. Sew the tufting pattern. Fold along the centerlines and sew along the sewing lines. If done properly, the sewing lines will be in line with each other. Lock-stitch or tie the endings of every sewn section to prevent the seams from loosening with service.

   ~Sew the perpendiculars first (those that go from the outside button spots to the edges of the panel). This makes it easier to establish the diagonal folds that will be done next.
   ~Sew the diagonal seams.

3. After the tufting seams have been completed, sew all of the panels for the complete pillow back together, except for the bottom where the filling material will be blown or stuffed in. After all stuffing is in, sew up the bottom. This can usually be done on the machine. Some, however, are sewn by hand. This back is filled with a combination of chopped foam and shredded cotton. The twine is looped in each button, ready to be attached to a recliner.

4. Install buttons through the back, attaching it to the unit base (FIG. 8-19). Because loose stuffing was used in this unit, the buttoning process was done with the unit lying on its back. This helped keep the filler from settling toward the bottom. Also, because a softer headrest and a firmer kidney roll were desired, buttoning was started at the top center and progressed toward the bottom (FIG. 8-19). This procedure tends to hold the softer stuffing in place at the top while pushing and packing the remaining

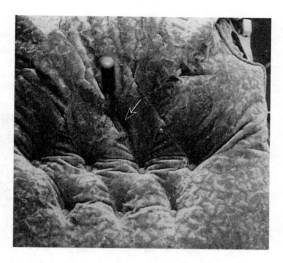

8-19 Applying another button with tufting needle (arrow points to button and short section of twine left on the face side).

filler downward to create a firmer bottom. Easy-chair comfort can thus be tailored in. The arrow in FIG. 8-19 points to one button being installed. Notice the white tying twine between the tufting needle and the button.

5. As each button is placed and snugged down, the next spot must be stretched out and the center precisely. Figure 8-20 shows a worker stretching one of the diagonals while locating the precise center with the forefinger. With the center located, the tufting needle, with both strands of twine threaded about 3 inches through the eye, is carefully placed, then forced through all padding materials and the foundation until the twine slips free from the needle eye. Snug up each button as it is placed, using the upholsterer's knot shown in chapter 7.

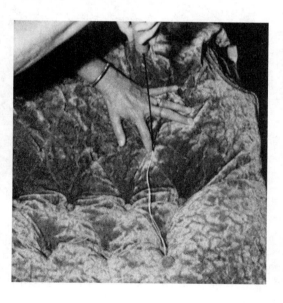

8-20 Stretching fabric and locating exact center of next tuft with finger while inserting tufting needle.

Tufting **127**

Figures 8-21 through 8-24 show what happens to fabric as tufts are sewn in. Figure 8-21 shows the back of a standard sewn tufting (diamond size of 4¼ × 7 inch). The front of the fabric is shown in FIG. 8-22. No buttons have been installed, yet the indentations for the buttons are very pronounced. The tucks in this set were sewn at ¼ inch only.

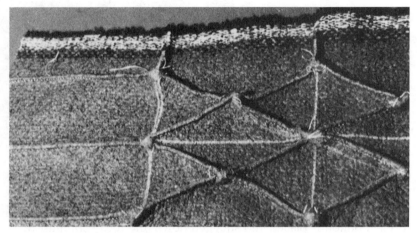

8-21 Close-up of the back side of a fabric with a sewn tufting.

8-22 Front side of sewn diamond-tuft pattern. Notice how the intersections are automatically recessed.

Figure 8-23 shows the beginning of a much smaller tufting—2¾ × 4 inches. In this style of sewn tufting, instead of sewing the tucks parallel to the centerlines, they were sewn in a curved, parabolic form. A closer look at the seam labeled D shows this curved tapering best. This curved sewing approach produces diamonds with inward-rounding lines rather than straight lines. This is a matter of personal choice. But as you might suspect from FIGS. 8-23 and 8-24, the smaller the size of the diamonds, the more difficult it is to sew and complete the tufting. With the smaller tufts (FIG. 8-24), small buttons must be used—number 22's.

8-23 Back side of smaller sewn diamond tuft (as the tufting size gets smaller, the sewing gets tougher).

8-24 Front view, showing the extent of "tucking" that takes place in smaller tufting jobs.

9
Finishing alternatives

In the past, period furniture designs could be identified by specific contours (camel back, Lawson, tuxedo, barrel, wing) and finishing techniques (Victorian, Colonial, Chippendale, Duncan Phyfe). Today, much of the furniture in the American home is a blend of the earlier purist features. Personal preference has created a permissiveness of features, and almost anything is acceptable. Each style or blend has its place, which is determined by the individual.

This chapter presents a few alternatives for finishing an upholstered work. Each has its specialty use, and each can be incorporated successfully to create interesting and pleasing effects.

BANDS

Bands are added to the tops and sides of arms and backs and to the bottoms of seats and outside panels to break up plain surfaces. Most bands are padded lightly, which gives an added touch of elegance. A few upholsterers have installed panels with no padding. This is an exception rather than the rule and is usually reserved for those units of a square, masculine appearance and the heavier-weight fabrics. Figure 9-1 shows a couch standing on its end with the seat-band panel laid back and the top portion of cotton padding stapled to the seat frame.

For furniture styles having the band attached to the top of the armrest, the inner-arm panel is tacked to the top of the arm, a welt stapled in place, and the band is blind-tacked using tacking strip, flexible metal, rigid metal, or cardboard. Figure 9-2 shows an arm band laid in place at the stump with the operator holding the remainder of the band panel back to permit stapling. (Normally, the full length of the band would be stapled in place with the entire panel in the full-open position.) Desired padding is added and the band folded over and stapled to the top arm rail along a line covered by the installation of the outside arm panel.

Keeping a band even in width and parallel to the lines of the unit is extremely important. Using a combination square set to a specified depth as illustrated in FIG. 9-3 is a convenient way to maintain the parallel when the edge of the frame can be used. The welt is brought up tight to the tip of the blade. The installation pictured in FIG. 9-2 is a band that extends from the front of one arm, up the back, across the top, and down to the opposite arm. The termination of the band at both sides is then concealed with panels when they are installed.

9-1 Couch (standing on end), showing seat band laid to one side and part of the cotton padding stapled in place. Also visible is the tacking strip holding the top of the seat band in place.

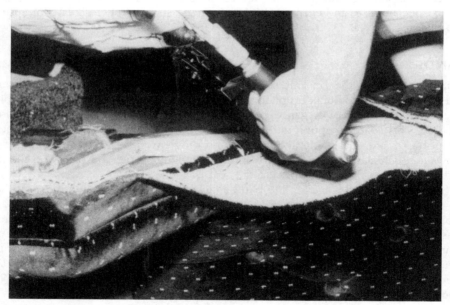

9-2 Stapling a tacking strip for an arm band.

WELT

Welt is used to finish off edges and junctions where two or more panels meet. Variations in the size of the welt cord create different impressions, from a delicate edging of a ⅟₁₆-inch cord to a bold border of a ½-inch fox edge. A double welt (see FIGS.

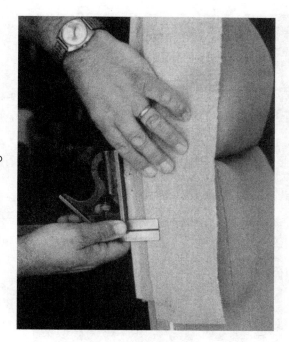

9-3 A convenient way to keep bands the same distance from edge of frame.

9-51 through 9-53) is a variation occasionally used in place of a single welt, as well as an edge finish much the same way decorative tacks or a gimp is used. Another popular use of the single welt is to finish off the bottoms of arms, slip seats, and the bottoms of chairs and couches. One such application is shown in the beginning stage in FIG. 9-4. The welt, attached along the front edge of the seat, will round the corner and be joined at the spot immediately in line with the front arm post. Joining welt ends behind a structural member is a good way to hide the joints completely.

Welt is normally located so that the outer surface is either flush with or protruding beyond the edge slightly (½ to ₁⁄₁₆ inch). In the application shown in FIG. 9-4, the welt has been left unattached about 2inches back from the corner to permit cutting a 90-degree notch (sketched in chalk) that will make a smooth, easy bend to form a square corner. The point of the notch should be at the same spot where the outside of the welt is to be located. The cut is rounded rather than a sharp V (FIG. 9-5). This makes a smooth yet square corner without bunching the fabric at the inside of the bend (FIG. 9-6).

Finishing a butt end

When welt is to butt against a post or other frame member and must "dead-end" into it, a completely finished end can be created in the following way:

1. Open the sewed welt for approximately 3 inches and cut the cord about ½ inch short of the end of the fabric (FIG. 9-7).
2. Cut straight into the fabric, toward the cord but ending about ₁⁄₁₆ inch short of the cord, as illustrated in FIG. 9-8. Make two parallel cuts as shown in FIG. 9-9.

9-4 Welt marked for a 90-degree cutout to form a square corner.

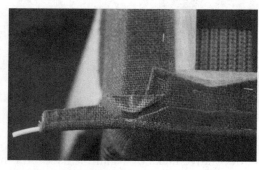

9-5 Notch cut out of welt, ready to form to staple.

9-6 Welt stapled to make square corner. (Raised edges, indicated by arrows, must be stapled down.)

3. On one side of the welt strip, cut off the inside corner so that the cut ends where the first cut (straight into the cord) ended. Then cut another wedge-shaped piece out of the opposite but *outside* the strip, as shown in FIG. 9-9 (notice the wedge lying on the seat). This will leave three "ears" extending from the end of the strip.

4. Fold the center ear so it lies directly over the cord (FIG. 9-10).

5. Fold down the other two ears so the end is square (FIG. 9-11).

6. Now fold the welt as it was originally sewn. (FIG. 9-12).

9-7 Cutting welt cord to make a "finished" butt end.

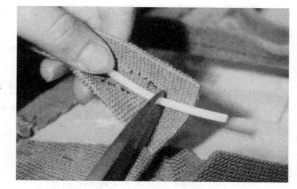

9-8 Cut welt strip within 1/16 inch of the cord, both sides of cord.

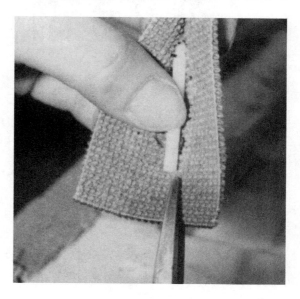

9-9 Cut wedges out of welt strip. One side is cut from the outside of the strip to the inner slit; the other side is cut from the inner slit to the outside of the welt strip.

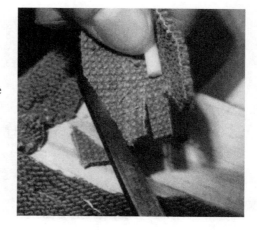

9-10 Fold over center flap to cover end of cord.

9-11 Fold over outer flaps to make a square end.

9-12 Close-up of a finished butt end.

The finished butt end can create an almost unnoticeable union of the welt with some fabrics when it completely surrounds an area and meets itself, such as a slip seat with no arm or back posts to hide weld joints. Figure 9-13 shows the beginning end being tacked to the bottom of a slip seat. Note the finished end and how the operator is using the finger as an edge guide to ensure an even projection of the welt beyond the edge of the seat. Notice also that the welt is being stapled tight against the cord. If this is not done, the welt will be loose and sloppy.

9-13 Stapling the first end of a finished butt joint to the frame. (Staple very tightly against the cord.)

To complete the perimeter, cut the ending welt so that it extends beyond the first about ½ inch (FIG. 9-14). Open the welt, make the cuts, and fold the end (shown earlier in FIGS. 9-7 through 9-12). Tack the second end tightly against the first, as shown in FIG. 9-15. A butt union, properly done, displays very little deviation, as can be seen in FIG. 9-16 (the arrow is pointing to the spot where the two ends meet).

Another form of the butt union is shown in FIG. 9-17. This union is formed by cutting the ends of the welt and cord flush and pressing the two together. This type of union works very well on only a few types of material; short-loop nylon frieze is one.

The lap joint

Another way to join welt is with the *lap joint*. The lap joint works well on fabrics that do not tend to fray at the ends. It is especially suited for vinyls. To make this joint:

1. Cut the cord and welt strip square and flush on the first end.
2. Determine where the joint is to be located (never on the front of a unit or cushion) and attach the welt around the perimeter, leaving 2 inches of the beginning end unattached and about 4 inches of the last end unattached.
3. Cut the finishing end square and about ½ inch beyond the beginning end (FIG. 9-18).

4. Open the finishing end of the welt about 2 inches and cut the cord so it ends at the face of the beginning end, (FIG. 9-19).
5. Wrap the finishing end tightly around the beginning end (FIG. 9-19).
6. Staple the union tightly to the frame (FIG. 9-20).

9-14 Cut welt ½ inch long to form the second end of a finished butt joint.

9-15 Stapling the second end for the butt joint.

9-16 Side view of a finished butt joint (junction is above arrow).

9-17 Making a plain butt joint.

9-18 First step in making a lapped joint.

9-19 Cut cord to end at face of first end, overlap second flap, and fold in as tightly as possible.

9-20 Completed lap joint.

BLIND STITCHING

One of the oldest of the finishing methods, *blind stitching* is still used for numerous functions. Among these are: (1) attaching new decking to an upholstered unit (FIGS. 14-21 through 14-25); (2) correcting an erroneous cut in a cover panel (which decreases as upholstering experience increases); and (3) closing pleats, folds, or tucks.

The blind stitch is a square stitch that is easiest sewn with a curved needle. The shorter the stitches, the more invisible (blind) the seam. Figure 9-21 shows a tucked corner of an attached cushion back in crushed velvet fabric. The tuck has been opened a bit to make it more perceptible. If the opening between the two layers of fabic is objectionable, close it with the blind stitch as follows:

1. Begin the stitch by inserting the curved needle into the fabric where the knot in the end of the thread can be hidden by the closed seam. The preferred location is near the junction where the overlap begins.

2. Exit the fabric a very short distance (⅛ inch or less) to the side of the beginning of the fold. Figure 9-22 shows these first two steps.

3. Snug the thread and hold it at 90 degrees (square) to the lay of the fold. Enter the fabric on the opposite side of the opened seam immediately beneath the thread and about ¹⁄₁₆ inch away. Exit the fabric at the outer edge of the fold, ³⁄₁₆ to ¼ inch away. The point of the needle can be seen at the tip of the arrow (FIG. 9-23). Tighten this stitch.

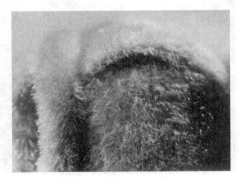

9-21 Close-up of tucked corner, showing slight opening.

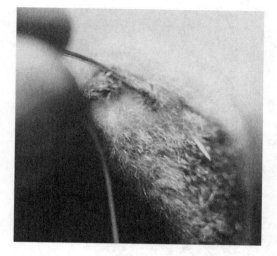

9-22 Starting the blind stitch to close up a tucked corner fold.

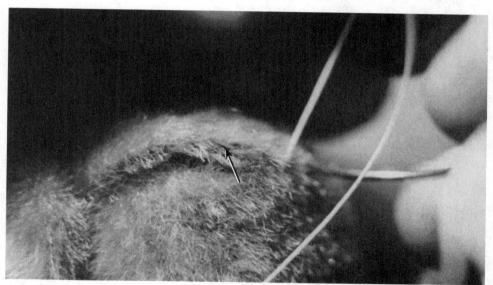

9-23 Make stitches short, and square to the lie of the fold.

4. Hold the thread square to the lay of the fold to locate the spot to enter and exit the opposite side, as described above. Figure 9-24 shows the beginning of the next-to-last closing stitch.

5. Complete the closing, the last stitch made so that the final knot in the thread can be hidden from view. Figure 9-25 shows the needle directed toward the small space between the welt and the inside back cover. By pulling the welt open, as shown in FIG. 9-26, the needle can exit the fabric and the ending knot tied and concealed where it will not be seen. The finished sewing is shown in FIG. 9-27.

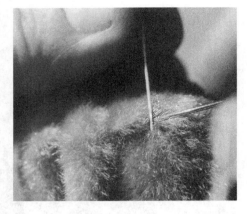

9-24 One method to keep stitches square to the fold line.

9-25 Selecting a way to hide the last of the blind stitches.

9-26 Forcing the welt open to hide the ending of the blind-stitch knot (yet to be tied).

PANELS

The use of *panels* is one of the most popular, practical, and pleasant finishing techniques. Their primary function is to conceal the unfinished edges of fabric tacked to the frame in very conspicuous locations (arm stumps and back posts). Figure 9-28 shows an operator checking a panel pattern (a piece of ⅛-inch hardboard) against

9-27 Completed blind stitch, closing a tucked corner. (No thread or knots are visible.)

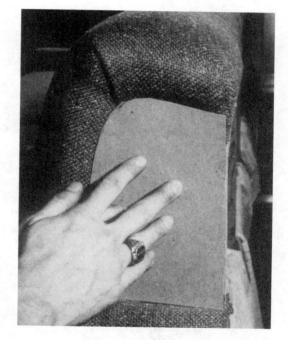

9-28 Checking the size of a panel pattern to a sofa back.

the back post of a sofa. The panel itself will be made of ¼-inch plywood, ⅛- or ¼-inch hardboard, or a heavy cardboard especially made for that purpose. Figure 9-29 shows a hardboard pattern and two different panels of ¼-inch plywood. The panel to the left is to be used on the front of an arm stump; the one on the right will be used for the back post, as illustrated by the pattern being checked in FIG. 9-28. To prepare panels, follow this procedure:

1. Bevel the top outer edges of the panel until they are about ⅛ inch thick. Make the taper about twice as wide as it is deep. Figure 9-30 shows the bevel more clearly than does FIG. 9-29.

2. Pad the face side of the panel. Either a full or half thickness of cotton can be used. Place the cotton on a flat, clean, smooth surface (like vinyl or Formica).

9-29 Beveled edges on ¼-inch plywood panels, back panel pattern is in center.

9-30 View showing more clearly the bevel on a ¼-inch plywood arm panel.

Lay the panel on top, the beveled side down, against the padding. Press the panel firmly into the padding and tear the cotton so it extends ⅛ inch to ¼ inch beyond the edges, as shown along the top of the operator's fingers, FIG. 9-31. A completed panel is shown at the very top of FIG. 9-31.

9-31 Tearing cotton padding for an arm panel. Finished panel shown to right.

3. Place the padded panel on top of a piece of cover fabric, the finish side of the fabric down, (FIG. 9-32). Fold the cover over snugly along the straight sides and staple to the back side of the wooden panel. *Caution:* Use short staples, ³⁄₁₆ or ¼ inch, for this purpose or they will protrude through the padding and cover on the face side.

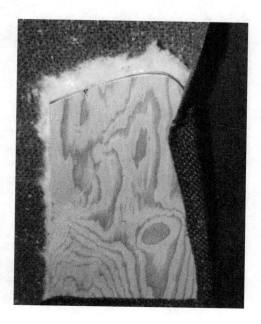

9-32 Bottom and right side of fabric has been stapled to panel base. (Keep the lines of the fabric pattern running true to the vertical line of the panel.)

4. For curved edges, place staples close (within ¼ inch) to the edge and trim off the excess as progress is made. This reduces bulk and prevents ripples on the front side (see FIG. 9-33).

5. Staple the sides before completing ends (FIG. 9-34). The chalk mark was made to indicate the curved edge of the wood. Notice also the diagonal placement of the staple (arrow). It is located in that position to facilitate the rounding of the corner. Cut the fabric alongside the staple to near the edge of the wood insert, as indicated by the dashed line.

6. Fold the cut tab under and pull the fabric into the corner as indicated in FIG. 9-35. Staple close to the edge (FIG. 9-36).

7. Work from both sides around the corners and finish off by pulling the center straight into the panel. Staple perpendicular to the edge and cut (FIG. 9-37).

8. Finish off the square end last by stapling the two corners diagonally, cutting toward the corners and folding the end over and stapling. Figure 9-38 shows one corner stapled and cut (with a bit too much cotton remaining). Simply remove the excess cotton and staple the end flap in place and trim. Occasionally, the square end is left unattached to the panel and pulled under and tacked to the bottom of the front seat rail. This creates a completely finished edge, preferred by some upholsterers.

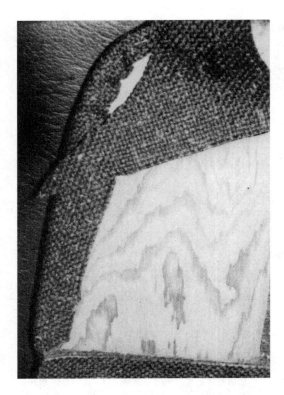

9-33 Forming around curves of a panel. Staple parallel and close (¼ inch) to the edge, trim excess, leave no ripples visible from front side.

9-34 Forming sharply rounded corners—staple at a diagonal (arrow), cut out excess (along black dashed line).

9-35 Tuck and form fabric tightly to the corner, pulling diagonally.

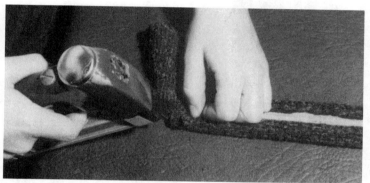

9-36 Staple close to edge and trim excess fabric before next tuck.

9-37 Cutting out last of excess fabric on a sharply rounded corner.

9. Attach the panel to the unit by driving 4-penny finish nails through the panel, fabric and all, from the front, as illustrated in FIG. 9-39. Drive the nails until they barely snug against the wooden insert. A dimple will be noticed in the fabric (arrow, FIG. 9-39).

9-38 Diagonally staple and trim fabric to form a square corner.

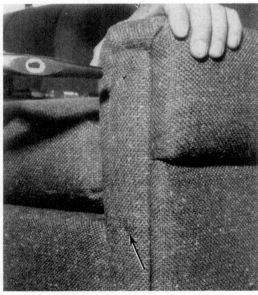

9-39 Attach completed panel with small finish nails (4d or 1-inch brads). Set nails snug (but not hard), as indicated by dimple at arrow.

10. "Lift" the fabric over the head of the nail with the point of an ice pick, as shown in FIGS. 9-40 and 9-41. Figure 9-42 shows the left arm of a chair finished with a panel, the other not yet installed.

9-40 Insert regulator along side nail head to "pop" fabric outward.

9-41 Cover fabric is pulled over head of finish nail.

9-42 Before (left) and after (right) view of arm-panel installation.

GIMP

Gimp can be used anywhere fabric butts against show wood. Figures 9-43 through 9-50 show the procedure for installing gimp with adhesive. (The adhesive should be the white polyvinyl variety or a special *fabric adhesive* that looks very much like the polyvinyls. These glues dry clear or transparent.)

1. Staple the cover close and parallel to the show wood, and trim off the excess fabric (FIG. 9-43).

2. Measure and cut the gimp to length. Allow about ½ inch extra at each end to be tucked under for a finished end, as indicated at the left side of FIG. 9-44.

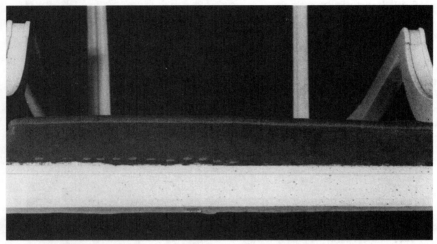

9-43 Front of chair with fabric stapled and trimmed in preparation for gimp.

9-44 Measuring gimp for length (leave about ¼ inch extra at each end).

3. Apply a ribbon of adhesive along the length to which the gimp is to be attached (FIG. 9-45).

4. Gently press the back side of the gimp onto the glued area (FIG. 9-46), making sure that your ½ inch still overlaps at the ends.

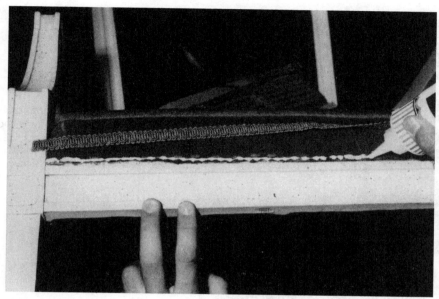

9-45 Applying adhesive to bond gimp (chair is lying on back).

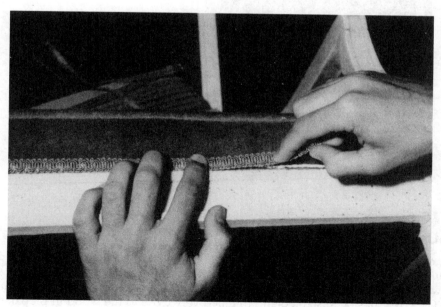

9-46 Press gimp evenly and smoothly into adhesive.

5. Apply adhesive to the extreme ends of the gimp and tuck in place with the point of a stuffing regulator, as shown in FIGS. 9-47, and 9-48.

6. Use the regulator point to make a nice straight edge against the show wood (FIG. 9-49). Figure 9-50 shows the completed application.

9-47 Apply extra adhesive to end of gimp to assure no fraying and secure bond when it is folded over.

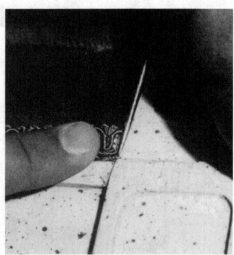

9-48 Tuck gimp under or into at corners.

DOUBLE WELT

The *double welt* was reserved for discussion at this point because it can be used in locations where single welt, decorative tacks, or gimp can be used. Figure 9-51 shows the outside of a wingback chair that has show wood all around the perimeter. The cover has been stapled and trimmed very close to the wood in preparation for the application of a double welt. Figure 9-52 illustrates a double welt being sewn

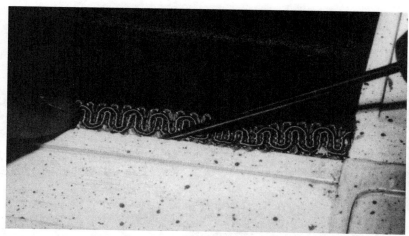

9-49 Regulating gimp along edge with ice pick (regulator).

9-50 Completed gimp finish.

with a special foot on the machine that makes the creation of a consistent, tight double welt much easier.

Application of the double welt is much the same as gimp, using either gimp staples or fabric adhesive. The installation around the carved leg, shown in FIG. 9-53, gives an idea of what the double welt looks like as a finishing method. This is the leg of the chair pictured in FIG. 9-51.

IRREGULAR CURVES

A variety of techniques can be used to finish off edges that fit against irregular curves. Figure 9-54 shows a modification of the Y cut (explained fully in chapter 13)

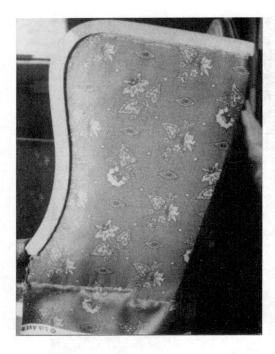

9-51 Outside of a wingback ready for a double welt to finish the edges.

9-52 Double welt being sewn with a special walking-foot attachment.

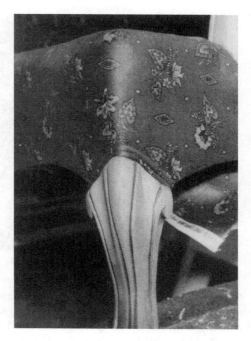

9-53 Double-welt application around irregular curves of chair leg.

to fit a seat cover around an exposed arm post. The center tab extends down the face of the post, and the flaps fold under and finish against the sides of the post, as with the Y cut. Notice that the two angular cuts, (FIG. 9-54), are not as wide, proportionately, as those discussed in chapter 13. You can see that the width of the Y cut is considerably narrower than the show wood it is to span, (FIG. 9-55).

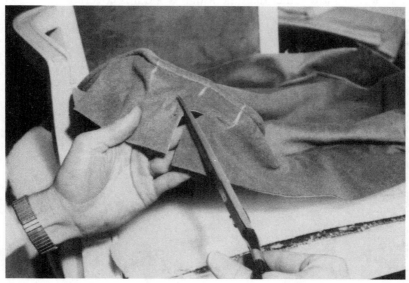

9-54 Modified Y cut for finishing around an arm post.

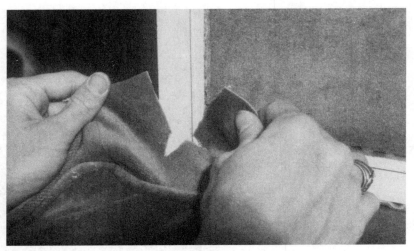

9-55 Opened Y cut to show how it will fit around arm post.

Another finishing technique is the installation of a *spanning panel*. This is used where two finished edges of fabric must span show wood and there is no other convenient way to cover the frame wood behind it, as in FIG. 9-56. The spanning panel is cut from a piece of cover to match the grain and pattern of the material that will face around the show wood. The spanning piece is then folded so that the rough edges will be against the frame, presenting a smooth, finished fold on the outer or visible edge. In this case, the flaps of the inside back will be folded under, pulled around the top and bottom of the arm, and stapled to the back upright. When the outside back is installed, only a small section of the spanning panel will be visible. But the fabric will show rather than the rough wood it is covering.

9-56 "Spanner strip" to cover frame wood at the rear of a show-wood arm.

TOP STITCHING

A couple of modern finishing techniques are created by a slight change in sewing technique. Figure 9-57 shows a mid-back panel stay tacked in place which has no welt around the seam. This weltless seam gives the unit a smooth, rather streamlined air. A slight modification to the weltless seam can be done by adding two top stitches, one along each side of the seam. This approach maintains the streamlined influence and adds elegance.

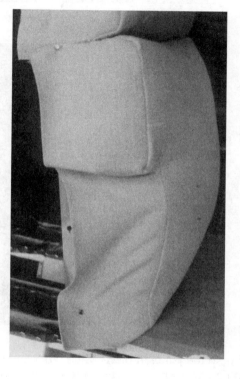

9-57 The plain seam used as a form of "modern" cushion finish.

DECORATIVE TACKS

Decorative tacks are used as a finishing technique in applications involving straight as well as irregular curving lines. They can be used in much the same way as gimp, but where the "knobby" effect of the raised tack heads is preferred to the more refined appearance provided by gimp. Irregular surfaces around arm posts are an especially appropriate use for the decorative tack. Finishing off around carved legs can be time-consuming. The decorative tack process is illustrated in FIGS. 9-58 through 9-61.

In FIG. 9-60, the seat panel has been pulled into place and tacked along the front and sides to within a few inches of the corners. The cover has been trimmed to over-lap the show wood around the legs by about ¼ inch. A portion of the panel has been lifted up (with a piece of white paper in the background) to show a short slit (just to the left of where the upholsterer's pin enters the fabric) cut into the cover to aid in tucking the material under to form a finished edge around the irregular curves of the leg. Notice that a stay tack (FIG. 9-59, just to the right of the finger) has been used

9-58 Slitting cover to accommodate tucks to fit around irregular curves.

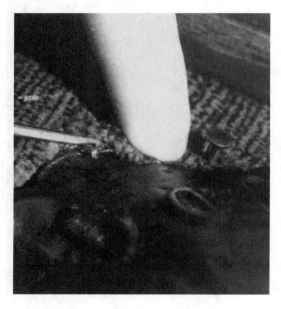

9-59 Using regulator to help form fabric to irregular curves.

to hold the panel in place while other cutting and fitting is being done. When all the fitting has been completed, a few decorative tacks strategically placed around the irregular perimeter (FIG. 9-60) will hold the cover permanently in place and lend that "antique" feeling appropriate to the carved leg (FIG. 9-61).

9-60 Decorative tacks as a finishing touch.

9-61 Decorative tacks finishing around carved legs.

10
Working
with vinyls

This chapter deals with vinyl covers for which no sewing is required. This technique is used frequently for covering dinette or kitchen chairs. Some padded bench seats are also covered with no sewing. This has special appeal to the hobbyist or homeowner who has no access to an industrial sewing machine.

STYLING

Since kitchen or dinette furnishings are more subject to food and beverage spills than the average living room furniture, easy cleanup is an advantage, especially when small children are still living in or visiting frequently. For this reason, a furniture style having smooth contours, no welts (the traditional dust and crumb catchers), and a minimum of tucks may be preferred. This style of reupholstering is known as *formed upholstery*.

Formed upholstery requires no sewing, eliminates the crumb catcher, and displays a rounded contour having small or no tucks or gathers on the seat and only unobtrusive tucks at the corners of the back. Forming is best achieved with the more stretchable vinyl fabrics.

DISASSEMBLY

Kitchen chairs are built differently than traditional overstuffed units. Therefore, a brief explanation of disassembly procedures will prove helpful. Generally, for convenience, speed, and comfort, it is suggested that the backs be removed before the seats. (There will be times when having the seat still attached makes it much easier to remove the back unit. This will become obvious in style 4, below.)

Backs

There are several basic systems for attaching backs.

1. Exposed steel frames with visible screws (from the back). Remove the screws. The back will usually release easily.
2. Exposed steel uprights, back inserted between them, screws not visible. This style is identified by caps covering the tops of the uprights. Remove the caps first; then look inside and you will see screws just a short way

down a groove on the inside portion of the tubular steel uprights. To remove the back:

~Remove the caps (or plugs) from the top of the uprights. This exposes the sliding screws. Remove one of the screws; then just loosen the other one.

~Pry one side of the frame sideways from the back near the bottom to expose a bottom pin. Continue wedging the two components apart until the pin releases from the hole in the frame. This takes some effort.

~Move the back toward the freed side to release the other pin. If this is not possible, wedge out the other side as before to release the second pin.

~With both pins free, slide the back upward and out of the grooves.

3. Exposed wooden frames. In this style, screws through the sides of the wood posts are sometimes concealed by wooden plugs or buttons. (Take care not to mar or lose the buttons; you will want to reuse them.) Carefully pry out the plugs or buttons with a sharp instrument, such as the point of a knife, an ice pick, an awl, or scribe. Take out the screws and lift the back upward.

4. Press-fit back, no exposed frame. This style has the uprights embedded within the back from the underside, which is actually the case. The back is removed by pulling it upward. One of the best ways to get if off is to kneel on the chair seat and pull up on the back, alternately rocking it back and forth until the back comes off. This is one style where it helps to have the seat still attached.

In some styles, the metal uprights have a burr or cleat that digs into the wooden portion of the back to keep the back from slipping off when the chair is lifted by the back. The retaining power of the cleats must be overcome to remove the back. This requires literally ripping the back upward, off the cleats.

Sometimes it is safer for your back to use a mallet and a block of wood to drive the back off the uprights. To do this, turn the chair upside down and rest the seat on a bench. (It is handy to have someone help hold the chair steady.) Place a block of wood against the bottom of the back, alongside one upright, and strike it sharply with the mallet (or hammer). Work alternately from side to side until the back drops off. Don't try to drive one side all the way off in one motion.

On other units, wooden dowels are pressed into the metal uprights, which in turn are pressed into the back unit. This style back is removed in the same manner. If pulling on the back does not release it, rather than straining your back, turn the chair upside down and proceed as directed in the preceding section.

Seats

Most seat assemblies for kitchen and dinette furniture are attached in basically the same manner—screws through the frame, visible from the underside of the unit. Remove the seat from the chair frame by turning the chair upside down and extracting the exposed screws. Lift the frame from the seat unit. Occasionally separation may require a little wedging or tapping pressure to overcome the fabric or seat base sticking to the frame. Raising the frame and tapping on the seat bottom will usually do the trick. Quite frequently the bottom of the seat unit is covered with cardboard. Save it if you can. This piece is put there to reduce or prevent squeaking between the frame and the wooden or metal seat base. If it is destroyed in the stripping process, just get some other lightweight cardboard or heavy paper and make new panels. This will be described later.

STRIPPING

The only real differences between stripping kitchen and overstuffed furniture are that the kitchen units are much simpler and the covered components are separate entities and are removed from the frame. Because the structure is slightly different, a short section on stripping is included.

Figure 10-1 shows a back (style 4, above) that has been removed. The outside back panel was put on with decorative tacks after a welt trim. This style is fast and easy. One serious disadvantage of this style is discovered when the chair is used against walls or other furniture. The tack heads scratch and mar the surfaces. Quick removal of the outside back and welt is accomplished by prying both the welt and the panel out, using a ripping tool.

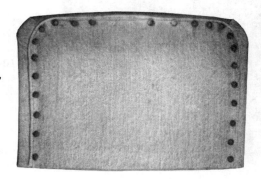

10-1 Rear view of press-fit back, outer back attached with decorative tacks.

A second style of back is shown in FIG. 10-2, where a cardboard panel is applied directly to the frame before the inside back panel is tacked on. The cardboard provides a smooth outer contour. With age and use, the cardboard may become creased at the edges of the uprights.

Another back style is shown in FIG. 10-3. The outside back panel is first glued to a cardboard base. This outside back assembly, then, is applied with decorative tacks.

PADDING & COVERING TECHNIQUES
Seat

Figure 10-4 shows a seat padding that was retained and will be covered with a new layer of cotton. The original padding consisted of a layer of foam rubber (V) and a couple of layers of cotton felt (W).

1. The entire seat assembly is placed upside down on top of an oversized piece of cotton and the excess trimmed to provide padding that will reach to the wooden edge (FIG. 10-5). Note how the worker is holding the cotton down with the finger tips while tearing around the perimeter of the base. Don't cut cotton; tear it. This gives a better feathering action.

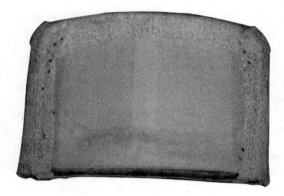

10-2 Outer back panel removed, revealing cardboard, and tacks holding inner back panel to the frame.

10-3 An outer back style, panel glued to cardboard base.

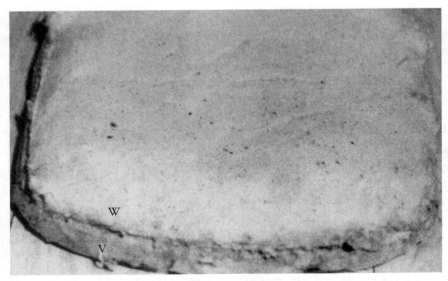

10-4 Stripped dinette chair seat showing that the padding is in good condition.

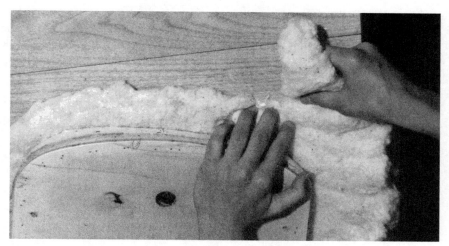

10-5 Tearing new cotton layer oversize to complement seat padding.

2. Place the completely padded seat assembly on top of the seat-cover panel. Tack the cover to the front center. Smooth and stretch vinyl fabric to the back, and tack the center in place. Smooth, stretch, and tack both sides at the centers (FIG. 10-6). *Caution:* Do not let the padding extend beyond the bottom edge of the wood as the cover is stretched into place. This is very important! A cover stapled with padding beneath it will result in an unattractive wavy surface that is impossible to smooth except by completely redoing it. Notice the tendency of the outer padding to creep

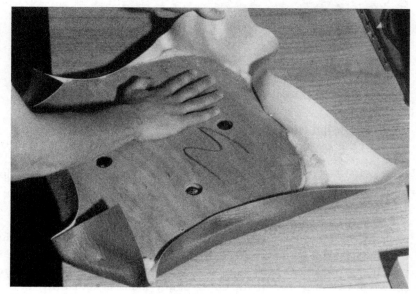

10-6 Applying seat cover. Tack centers first, then corners.

beyond the edge in FIG. 10-6. Slide the stretched fabric outward over the edge of the wood while applying a little pressure with the hand from the outside. Then, retaining that pressure over the cover, pull the fabric back over the edge, and the padding no longer extends beyond the edge but forms a nice roll just at the edge, with no lumps, bumps, or ripples. The upholsterer has just done that at one corner in FIG. 10-6.

3. Stretch and fold corners at a 45-degree angle. Staple at the center of each one. Note that the third corner is being stretched in FIG. 10-6. Note also that there is no padding coming *beyond* the underside of the wood at the tacking point.

4. Form corners. Beginning near the center of the rounded corners, stretch cover at an angle toward the center staple to make a small tuck, on the bottom only. No wrinkle or fold should be seen at the seat edge. Staple across the tucks on a diagonal to hold them in place without slippage (FIG. 10-7).

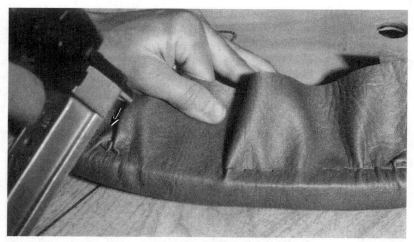

10-7 Taking small tucks to remove excess fabric at corners.

5. Continue small tucks around the corner to remove all excess fabric between the last corner staple and the adjacent center staple (FIG. 10-8). Staples are very close together, and the tucks are very small. This, and the stretchability of vinyl, is what prevents wrinkles. Continue this procedure for all corners.

6. Trim excess fabric ⅛ to ¼ inch from the staples to prepare for the bottom cardboard. Replace antisqueak cover (FIG. 10-9). Breather holes are provided in the base as well as in the cardboard—an absolute necessity to avoid entrapping air and an embarrassing sound when you sit down.

7. Attach antisqueak cover by stapling close to the edges of the cardboard, as shown in FIG. 10-10. Secure the edges of any small tears that may have been created in the stripping process.

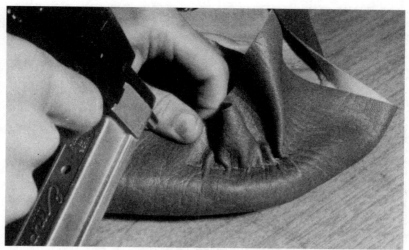

10-8 Corner tucks completed, excess fabric taken up in tucks, ready to tack toward center staples.

10-9 Applying anti-squeak cover to bottom of trimmed, formed seat.

Before attaching the seat to the frame, rub some paraffin wax along the cardboard in line with where the chair frame will attach. This reduces friction and squeak.

The small ripples visible in FIGS. 10-7, 10-9, and 10-10 can be removed with a heat gun. Just heat the vinyl to soften it and lightly rub it to smooth the ripples. *Caution:* Heated vinyl can become too hot to touch with comfort. Wearing an inexpensive pair of cotton gloves to smooth the heated vinyl avoids unnecessary pain.

10-10 Staple parallel and close to edge of antisqueak cover.

Padding the inside back

If more padding is desired, add another layer. Smaller pieces of cotton can be used for padding if each one is "feathered" along the line where the two will join. Tear half the cotton thickness away to about 1½ inches back from the edge (FIG. 10-11). Feather the joining piece in similar fashion and overlap the two feathered portions as shown in FIG. 10-12. If done with care, this creates a union in which gaps and lumps are undetectable and the pieces joined do not slip apart. For best results, have the cotton seam run up and down the back and seat rather than from side to side. This reduces any tendency to have the seam separate because of sliding action.

10-11 Preparing to splice a new layer of cotton padding.

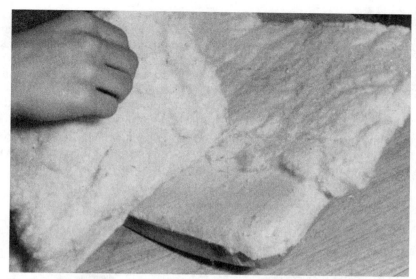

10-12 Joining feathered cotton pieces for a smooth splice.

Covering the inside back

1. Attach top center. Staple the top center of the inner back cover to the back at a point that will be concealed when the outer back panel is put on. That first staple is not quite visible at the extreme top of FIG. 10-13. Pull the center of the fabric snugly over and around the top edge and down toward the bottom.

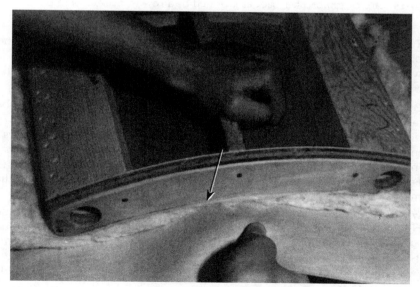

10-13 Pulling bottom center to begin forming IB panel to frame.

2. Tack bottom center. Stretch fabric quite tightly to bottom center. If the center is not stretched tightly, significant bridging will occur later when the vinyl is stretched toward the sides, and the inside curvature of the back will be lost. Notice that the cotton goes to the front edge (FIG. 10-13, arrow) but does not go beyond the edge onto the bottom surface. The cotton visible toward the outer edges of FIG. 10-13 will be tucked or rolled to meet at the edge but not beyond it. Ripples will form at the bottom and top, but they will be removed as the "forming" process continues.

3. Forming the top and bottom. Pull the top edge at a slight diagonal toward the corners (FIG. 10-14) and staple down, working from the center to within 4 inches of the corner. Follow the same procedure to attach bottom.

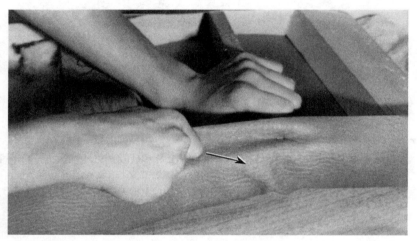

10-14 Stretch top of inner back as indicated by arrow to make a wrinkle-free top surface.

4. Tack side centers in place. Gently smooth (do not stretch tightly) the inner back cover to the side centers, and tack. Then form and tack sides to within 2 inches of each corner, top and bottom, using the same diagonal pulling as in step 3.

5. Forming top inner back corners. There are several alternatives that can be taken when forming corners. The sharpness of the corner radius often dictates which alternative is best. For corners having a radius of 2 inches or more, the smooth rounding works well (FIG. 10-15). For corner radii of ½ inch to 2 inches, try the multiple tuck (notice the double tuck in FIG. 10-19). For sharp corners or those having a radius less than ½ inch, one major tuck may be the choice.

Alternative 1: Smooth, rounded top corners. As with the seat, take small tucks on the back side, and staple. To create a smooth, rounded corner, take small tucks around the corner. (The small wrinkles beneath the joints of the thumb in FIG. 10-15 should be removed by taking out the last two staples and restretching). Create the smooth rounding by pulling the fabric around the top of the corner, as indicated by the arrow in FIG. 10-15.

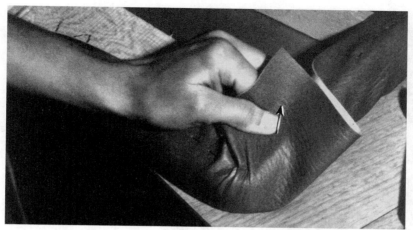

10-15 Forming top corners of inner back panel.

Alternative 2: Major tuck at corners. To form this type of corner, stretch the fabric to the side to take up the slack and create the major tuck location that is attractive to you. Experiment a little at this point. The student is working with alternatives in FIGS. 10-15 and 10-16. Notice that trying to make a single major tuck in a corner with a radius greater than 1 inch does not work easily (FIG. 10-16). When the slack has been taken out of the sides, eliminate excess bulk from the tuck by cutting out excess material. To do this, hold the fold tightly and open to expose the underside. Cut on a diagonal from just inside the last corner staple to within ¼ inch of the fold, indicated by the tip of the shears in FIG. 10-17. Now cut ¼ inch away from the fold line to angle into the ending point of the previous cut (FIG. 10-18). Do not let the fold shift when making this cut. This procedure removes excess bulk from any tucking operation and gives a smooth, flat tuck (FIG. 10-19).

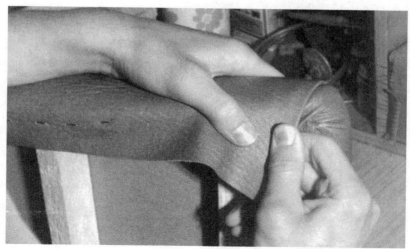

10-16 Establishing major tuck for top corners.

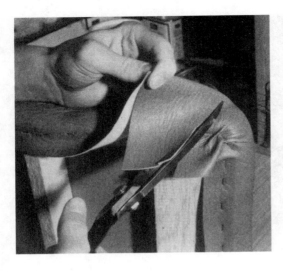

10-17 Maintaining fold line while making first cut to remove excess fabric at corner.

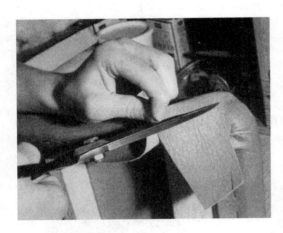

10-18 Cut within ½ inch of fold line to remove excess fabric for tuck.

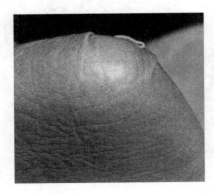

10-19 Front view of "formed" top inner back corner. (A tab of the outside back is also visible.)

Alternative 3: Multiple tucks. Making more than one tuck for corner radii of 2 inches or more is acceptable. However, tucks must be facing outward and downward and be located at the same spots on both sides of all chairs of the set. Figure 10-19 shows a small side tuck facing downward with the top tuck facing outward, as they should be. Never make a side tuck face upward, or a top tuck face inward.

Covering outside back

One of the most popular methods of attaching outside back panels is blind tacking. The procedure is basically the same as with overstuffed upholstery styles explained in chapter 16.

1. Locate top and bottom outer back centers. Mark the top and bottom centers of the fabric (with a V notch) and the frame (with a straight line, FIG. 10-20). The outer back cardboard is already in place.

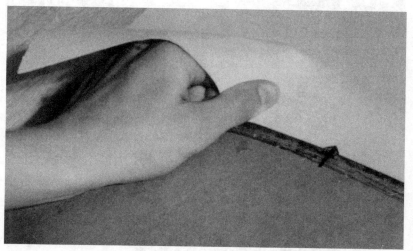

10-20 Locate top center of outer back panel to center mark of frame.

2. Staple center and stay-tack top. Staple the center of the fabric at the back center and stay-tack fabric toward the sides at approximately 3-inch intervals. Allow ½ to ¾ inch for blind tacking, as shown in FIG. 10-21. This is merely to hold the fabric in place in preparation for the tacking strip.

3. Center and attach tacking strip. Attach a lightweight cardboard tacking strip along the top so the top edge of the tacking strip is ⅛ to ¼ inch from the frame edge (FIG. 10-22). Staples in the tacking strip should be quite close together and near the edge, as shown in FIG. 10-22.

4. Pad the outside back. Frequently it is desirable to lightly pad the outside back. To do this, use half thickness of cotton or ¼- or ½-inch polyfoam. Place padding to the tacking strip; staple 1½ inches to 2 inches apart, just to hold in place, then trim ½ inch inside a line where the inside edge of the side tack strip will be located (FIG. 10-23).

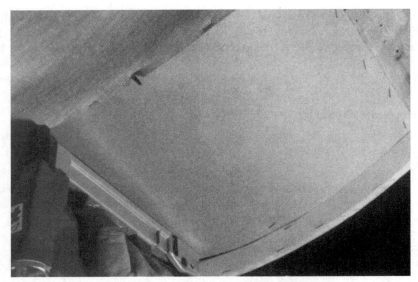

10-21 Lightly stretching and place-tacking outer back for attaching tacking strip (lightweight is best when working with vinyls).

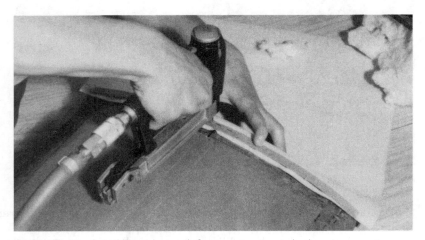

10-22 To attach tacking strip, work from center toward edges.

5. Attach bottom of outside back. Smooth and lightly stretch the outside back panel down and staple, progressing from center toward both sides (FIG. 10-24). You will notice that the vinyl has been folded under to give a finished edge. Leave 2 or 3 inches from each side unattended, as shown.

6. Apply side tack strips. Cut two lengths of tack strip for the sides. They should be ¼ to ½ inch shorter than the distance from the top fold to the bottom edge of the back. Place the inner edge of the tack strip along a line where the outer edge of the outside back is to be folded over (FIG. 10-25).

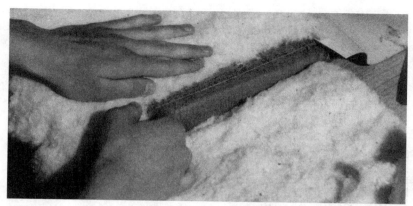

10-23 Tear cotton padding ¾ inch short of sides where tack strip is to be attached.

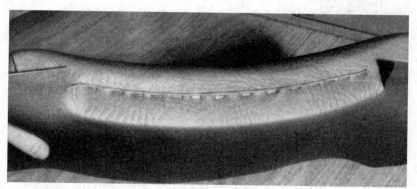

10-24 Center portion of OB panel stapled in place.

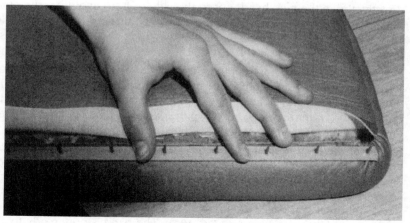

10-25 Place tack strip along sides of back with inside edge of tack strip along line where the fabric will be rolled under.

7. Press tacks through fabric. Lightly pull (diagonally outward and toward bottom) outside back panel over tack strip and press onto the tacks. Be careful. Those tacks are sharp. *Caution:* Press fabric onto tacks in two steps: (1) partially, to orient the fabric (FIG. 10-26), then (2) completely seated. Do not seat fabric completely until all tacks have been located and partially penetrating the fabric. Excess fabric should be trimmed ½ to ¾ inch from the tack line. Once the fabric is completely seated, roll the tack strip and fabric under.

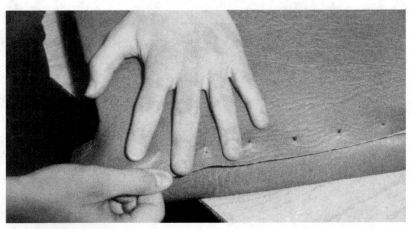

10-26 Lightly pulling outer back panel to set onto tack strip prior to turning under. Set all tacks part way at first, then go back and set fabric all the way to the base of the strip.

8. Partially seat tack strip. Stretching the fabric slightly to the side with one hand, begin at the top and tap the tacks partially into the frame with a broad-faced hammer or the side of a tack hammer. Hit the tack strip directly above the tack head, not to the side of it, to avoid bending tack, destroying the tack strip, forcing the tack head through strip and fabric, or all three. Do not drive the tacks all the way in at first. That could cause the problems mentioned.

9. Drive tack strip flush. Now, set the tacks all the way in as pictured in FIG. 10-27. Be sure to hit the tack strip directly above and straight onto the heads of the tacks rather than just anywhere along the tack strip. Placing a folded scrap of vinyl fabric over the outside back panel when seating the tack strip reduces the potential to mar, scuff, or cut the outside back.

10. Finish off the bottom as illustrated in FIG. 10-28. In this case, the fabric covering the holes for the frame uprights is slit from the center toward the edges. This procedure permits the fabric to fold inward when the uprights are inserted and serves to tighten the fit. In some cases, where the fit may be so tight that additional material would be unwanted, cut a complete circle in the fabric, as shown in FIG. 10-29. The back is now ready to slip back onto the upright posts of the frame.

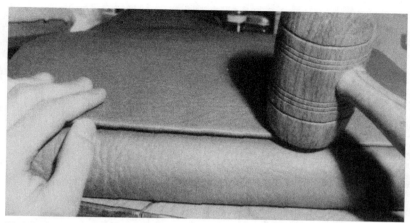

10-27 Setting tack strips tight to the back frame.

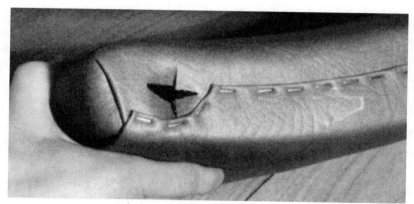

10-28 Bottom view of completed outer back installation, showing the cross slit that will make a press fit tighter.

10-29 A second style of back, showing nontightening cutout. This style has the outer back panel glued to a cardboard base, then attached to the frame with button-head tacks.

11
Industrial sewing machine

Sewing is almost as much a part of an upholsterer's activities as stripping and fitting. Most units require at least some sewing of welt and cushions. Numerous styles require extensive sewing. One example is the barrel-back chair; virtually all panels are sewn together before being fitted to the unit. Another style requiring extensive sewing is the recliner-rocker, which requires between 9 and 13 panels to be sewn together for the back attached-cushion assembly alone. Such extensive use of the industrial sewing machine merits instruction on its use, maintenance, and basic adjustments.

Only the straight-stitch walking-foot machine is discussed since the most popular machine for furniture upholstery. The photos in this chapter are of Pfaff, Model 145-H3 and Model 145-P. The principles of threading, tensioning, and sewing on the Pfaff models relate almost directly to any straight-stitch industrial sewing machine and are fundamentally the same for straight-stitch domestic models.

Some of the primary differences between the industrial and domestic machines, which will be immediately noticeable to one who has used a domestic machine:

- The industrial machine has noticeably more power, being driven by a much larger motor.

- The upholstering machine has a walking foot, as opposed to the stationary or sliding foot on most domestics.

- The foot treadle has three functional positions: (1) neutral, achieved by slightly depressing the toe of your foot on the rear of the treadle, but not enough to engage the power drive (neutral is especially useful to permit turning the handwheel for timing and other adjustments); (2) variable-speed power drive, achieved by depressing the toe beyond the neutral position (the further the toe is depressed, the more power delivered to the motor, hence increasing sewing speed); (3) brake, applied by depressing the heel of the foot on the front of the treadle (this applies a frictional brake to the drive system, instantly stopping the sewing motion).

- The industrial machine will "coast" to a stop when the treadle is released unless the brake is activated.

- The very sound of the industrial machine almost unnerves some who have experienced only the domestic models or are naturally timorous around machinery.

THREADING THE NEEDLE

The primary concern in threading a machine is to assure a continuous, unobstructed path for thread travel. The first requirement is to have the thread that goes to the needle come off the spool with no friction. This is accomplished by having it pull directly upward from a stationary spool. Figure 11-1 shows one of two spools of thread on one type of standard thread stand. The thread must go from the spool *upward* through the guide. Please recognize that the purpose of any thread guide is to keep the thread out of the way and moving freely.

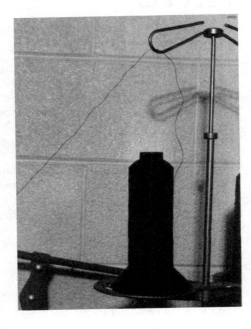

11-1 Spool of thread to feed to needle. (Upper thread guide is used to discharge thread straight up. Spool must not be required to turn to discharge thread.)

Figure 11-2 shows the basic threading path for the needle thread. Thread the machine in this order:

1. Guide post on top of the machine. It is there to keep the thread out of the working mechanisms to the top right portion of the machine. In that post there are usually two holes, drilled at 90 degrees to each other. The thread should pass through both of them. Some operators will use only one. The only difference in using one or both holes is that with two there seems to be just enough "drag" to give a smoother thread travel.

2. Needle-tension guide and roller. Pass the thread through the hole in the plate on the right side (FIG. 11-3A); over and between the rollers (FIG. 11-3B); to the right, around, and snugged between the polished disks of

11-2 Threading path from spool guide to needle: (A) thread guides; (B) needle-thread tension assembly; (C) overarm take-up assembly; (D) overarm; (E) pressure-foot height adjustment; (F) pressure-foot spring, (G) pressure-foot tension adjustment; (H) fulcrum for pressure foot; (K) attachment point of spring to pressure-foot bar.

11-3 Needle-thread tension and take-up assemblies: (A) thread guide; (B) tension roller guide; (C) tension disks; (D) overarm take-up assembly.

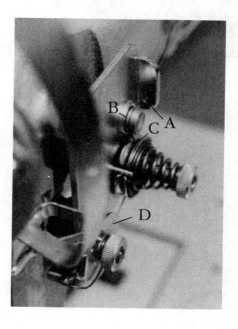

the thread tension adjustment (FIG. 11-3C); and straight off the bottom of the tension assembly to the bottom of the thread take-up spring assembly (FIG. 11-3D).

3. Thread take-up assembly. Hook thread into the lock at top of thread take-up assembly. This action will require using the right hand to hold the thread from moving (between the top post and the needle-tension guide is a good spot) while pulling upward with the left hand on the free end to move the spring and thread past the lock, (FIG. 11-4). Moving the free end slightly toward you and relaxing the tension with the left hand will permit the spring to take the thread downward from the hook to its natural position, (FIG. 11-5, right side).

11-4 Locking thread into the take-up hook.

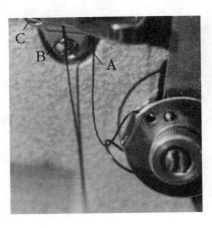

11-5 Take-up hook loaded properly and spring in relaxed position. Notice that the thread loops over hook, goes down through spring loop, up through guide (A), through overarm (not shown), and back down through opposite side of guide (B).

4. Overarm guide, overarm, and overarm guide again. Pass thread upward through the right side of the overarm guide (FIG. 11-5A); through the hole in the overarm, and back down through the left side of the overarm guide (FIG. 11-5B). This can usually be done by holding the thread taut between the fingers and passing it beneath the overarm guide hook (FIG. 11-5C).

5. Needle bar guide. Hook thread into slot on bottom left side of the needle bar (FIG. 11-6).

11-6 Needle bar thread guide. Loading is easier when holding thread taut and sliding into groove.

6. Needle. The last step is threading the needle. Always thread from the outside toward the inside of the machine. Figure. 11-7 shows the needle in its upmost position and the pressure foot down to permit easy access to the eye of the needle. The completely threaded needle is shown in FIG. 11-8. Notice that the thread is going through the needle from the outside (left, in this case) to the inside.

11-7 Threading needle; always thread from outside toward the inside.

THREADING THE BOBBIN

1. Open the bobbin cover. Depress the spring bar on the base of the machine to the right of the needle and slide the bobbin cover plate to the right. This will reveal the bobbin case, bobbin cap, and needle-pickassembly (FIG. 11-9).

11-8 Needle and bobbin threaded properly. Pressure foot is down, bobbin cover open slightly—not a good idea.

11-9 View of bobbin cap and case; bobbin-cap lock is in locked position.

2. Lift the bobbin cap lock (FIG. 11-10).

3. Remove the bobbin cap from the post (FIG. 11-11).

4. Replace empty bobbin with a full one. Be sure that the thread comes off the bobbin so that when pulling on the thread it will hold itself into the slot (C) and tension spring (B) in the bobbin cap (FIG. 11-12). Holding tension on the bottom of the bobbin with the left hand, pull the thread into the slot and beneath the spring. The operator is beginning this step in FIG. 11-12.

11-10 Bobbin-cap lock released; ready to remove cap and bobbin.

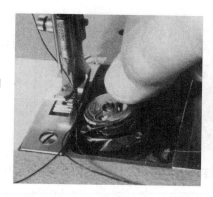

11-11 Removing bobbin cap and bobbin (bobbin still has thread).

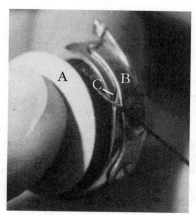

11-12 Loading bobbin into cap: the thread must come off the bobbin (A) from the left to enter groove (C) and be pulled beneath tension spring (B).

5. Lock bobbin thread in cap groove (FIG. 11-13.)

6. Replace newly loaded bobbin cap into bobbin case and lock in place (FIG. 11-14). Be sure that the thread will pass with relative ease through the small opening created by the grooves in the bobbin cap and case (illustrated A, FIG. 11-14).

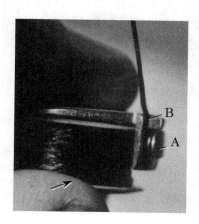

11-13 Face of bobbin cap: thread is beneath tension spring (A) and into cap notch (B). If properly loaded, bobbin will turn in direction of arrow as thread is pulled from cap notch.

11-14 Bobbin and cap loaded with thread properly exiting through cap notch (A). Needle thread (B) is shown coming around bobbin to pick up bobbin thread (C).

ADJUSTMENTS

If the sewing threads (needle and bobbin) are not interlocking near the center of the fabric being sewn, something is obviously out of adjustment. Any change in thread size, either for the needle or the bobbin, will upset the fine tension adjustments to either. It is therefore necessary for the operator to know how to make two basic thread-tension adjustments.

Needle thread tension

Tension adjustment for the needle thread is found on the front of the machine head (refer to FIG. 11-2B). Screwing the thumb wheel *in* (clockwise) increases tension on the thread; *out* (counterclockwise) reduces it. (If the pressure foot is up, a cam at the rear of the machine activates a pushrod that releases the thread tension. This permits easy pulling of the thread at the end of a stitch.) Adjusting the needle tension is preferred because it is easier and not as sensitive as the bobbin-tension adjustment (shown in FIG. 11-15).

11-15 Bobbin-thread tension adjustment. *Caution:* This adjustment is extremely sensitive when it gets near proper adjustment.

If the two adjustments are set properly, the interlocking loops of the needle and bobbin threads will meet (loop around) each other somewhere in the materials being sewn. Figure 11-16 illustrates a stitch with proper thread tension; no loops can be seen on the top or on the bottom of the stitch. If it can be avoided, make no adjustment on the bobbin tension, it becomes very sensitive when it is close to the proper setting.

11-16 Stitch showing proper needle and bobbin tensions.

Figure 11-17 shows a stitch in which the needle-thread tension is too tight. Notice that the interlocking loops of the needle and bobbin threads are visible on the top side. To correct this condition, decrease tension on the needle thread. If the needle thread tension is already loose, tighten it just a tad and tighten the bobbin tension.

11-17 Results of needle tension too tight.

Bobbin-thread tension

Figure 11-18 depicts a stitch with excessive bobbin tension. The interlock is now on the bottom side of the fabric. To correct, first try increasing needle-thread tension. If the condition cannot be corrected before the tension gets so great that it breaks the needle thread, reduce the needle-thread tension considerably, then reduce the bobbin tension by turning the screw *slightly* counterclockwise (FIG. 11-15). *Caution:* This adjustment is extremely sensitive if the bobbin spring is still in good condition. A small fraction of a turn of the screw makes a significant difference in the tension of the thread. Check the stitch. If the bobbin tension is still too great, loosen it. If it seems too loose, try reducing the needle-thread tension first. Work back and forth until optimum tension is achieved.

Important: The needle- and the bobbin-thread tensions are interactive. If significant changes in either tension adjustment must be made, the two should be balanced by working alternately with each to achieve the proper adjustment. Changing one affects the action of the other. Changing thread size also affects tension.

Industrial sewing machine **187**

11-18 Stitch with bobbin tension too tight.

Timing

The most frequently needed major adjustment for the upholstery sewing machine is the timing adjustment. If the bobbin pick isn't picking up the needle thread—in other words, if the machine isn't sewing—there are two major causes. Either the needle pick-drive dog has been "popped," or the timing is off. The drive dog will pop if the pick experiences an excessive drag in its rotation. Such drag is most often caused by the thread double-locking around the bobbin case or getting caught beneath it. The timing can be knocked out of adjustment by hitting something hard with the needle (like a zipper) or the needle pick rotation being obstructed. To check the timing, the bobbin cover must be opened and the foot plate removed (FIG. 11-19).

11-19 Removing foot plate to check timing. Notice that the second plate screw (behind foot) has been removed.

Two handy tools for working on the machine are the modified screwdrivers shown in FIG. 11-20. The first is a small offset screwdriver, straight-slot. This can be purchased, or a standard small screwdriver can be bent at 90 degrees (see bottom, FIG. 11-20). The second tool is a stubby straight-slot screwdriver. It can be purchased, or the handle of a medium-size driver can be cut off and filed smooth, like the one shown on the top of FIG. 11-20.

Timing check

The correct timing can be identified by the relative position of the needle pick to the needle. To determine this position, turn the machine in the forward direction with

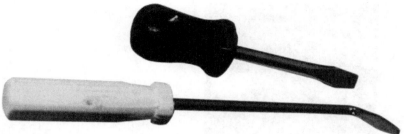

11-20 Two handy screwdriver modifications for repairs.

the handwheel until the pick is just approaching the needle from the front side of the machine. If you go beyond the desired point to make this check, FIG. 11-21, continue turning the machine in the forward direction to bring the pick up to the needle from the front. Do not turn the machine backwards. Backward motion may result in an inaccurate timing check.

The pick should be reaching the front edge of the needle with the point of the pick about ¹⁄₁₆ inch above the top of the needle's eye, with the needle traveling upward. If the pick position is correct but the needle is traveling downward, the timing is way off.

11-21 Setting as shown is in proper timing. Thread pick (B) is approaching needle just above eye (A) with the needle in motion upward (shown by arrow).

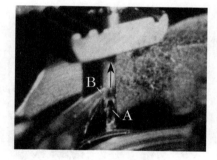

Timing adjustment, needle bar

Sometimes the timing can be adjusted by moving the needle bar up or down, as required. *Warning:* This is a limited adjustment. It requires coordination with the pressure foot, which will be specified later. To make this adjustment, first remove the head cover plate on the left side of the machine. The two screws, identified by arrows in FIG. 11-22, need only be loosened, not removed. Pivot the plate toward the front, then raise it upward, as indicated by the arrow sequence in FIG. 11-23. This will permit the plate to pivot out of the bottom screw and slide over the head of the top screw.

Loosen the needle-bar setscrew, identified by the tip of the screwdriver in FIG. 11-24. *Caution:* This screw must be tight; therefore, take care not to let the screwdriver slip and destroy the slot. Considerable torque may be required to loosen it. Loosen it only enough to permit the needle bar to be moved with a little pressure. If it is loosened too much, the bar will drop downward, further upsetting the timing ad-

11-22 Head cover plate: arrows indicate two screws to loosen for plate removal.

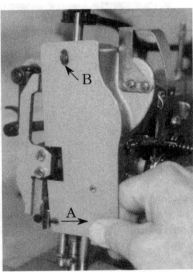

11-23 Removing head plate: rotate bottom forward (A); lift upward (B) to position enlarged section of groove beneath screw head, then pull plate straight out.

justment. Move the needle bar to establish the needle position indicated in FIG. 11-21, then tighten the setscrew.

The needle bar cannot be lowered to the point where the bottom of the bar will contact the walking foot when it is in the raised position. Figure 11-25 shows the needle bar in as close to the walking foot as can be tolerated. Note that the pressure foot is in the raised position and the needle is full down. Thus, if the needle bar has

11-24 Needle-bar set screw (for timing adjustment). Loosen to raise or lower needle bar. Screw must be very tight when operating machine.

11-25 Needle bar to pressure-foot adjustment. Bar is in full-down position, foot raised to full-up position.

to be lowered to a point beyond the position shown, the needle-bar adjustment cannot be used.

Timing adjustment, thread-picker gear

The more dependable timing adjustment is made by altering the position of the bevel gears located beneath the bobbin assembly. To get to these gears, tilt the machine head backward on its hinges to reveal the underside of the machine (FIG. 11-26).

Another source of sewing malfunction is depicted in this photo. The drive dog is out of its slot (FIG. 11-26A). This problem is easy to remedy and will be shown a bit later in detail.

11-26 View of underside of machine: drive dog and groove (points A); thread picker bevel gear cover (B).

Point B, FIG. 11-26, is the cover to the thread-picker bevel gears. To get to the gears, remove the housing screw (FIG. 11-27). The operator's hand is supporting the back side to avoid closing the machine head (sometimes it takes a bit of pressure to loosen the screw). Remove the cover (FIG. 11-28), and you can see the gears.

The gear to be moved is the smaller of the two. There are two setscrews in this gear, one of them visible in FIG. 11-29. Loosen the first one, then tighten it again until the screw snugs lightly against the shaft. Rotate the handwheel forward to expose the second screw; loosen and snug this one also. Now position the point of the

11-27 To access bevel-gear adjusting screws, remove cover screw supporting unit with other hand to avoid closing machine from pressure applied to screw.

11-28 Removing bevel-gear cover.

11-29 Bevel-gear adjustment: loosen, then lightly snug *both* set screws. One is located on each side of bevel gear.

thread picker just at the front portion of the needle, as indicated in FIG. 11-21. Holding the picker and bobbin assembly firmly with one hand, carefully and slowly rotate the handwheel until the needle is traveling upward and in the same position relative to the needle pick, as indicated in FIG. 11-21. This adjustment is rather touchy. A seemingly insignificant deviation up or down will cause the machine to skip stitches or fail to pick up the bobbin thread.

Tighten the setscrews, semifirm, on the bevel gear and without replacing the gear cover, try sewing a couple of layers of upholstery fabric. If the adjustment is proper, the stitch will be even and complete. If skipping or nonsewing occurs, try again. If the needle bar touches the raised pressure foot, raise the needle bar slightly, tighten the setscrew firmly, and reset the timing with the gear adjustment. When the timing is right, tighten all screws firmly, replace covers, and go to work.

Pressure-foot height adjustment

For best results, the walking foot must raise high enough to "walk over" the fabric without dragging it. As sewing thicknesses change, so must the height of the walking foot. Although this adjustment is relatively easy to make, it becomes bothersome to change it for every function; thus, a "general" position is most often selected and left there except for unusual cases.

Figure 11-30 shows the rear of the machine. Near the center of the photo you can see a wing nut just to the left of a darker bracket. This is our adjustment. Loosening the wing nut, moving the bolt all the way to the top of the slide, and retightening it, as shown in FIG. 11-31, will raise the walking foot to its maximum height, approximately 3/16 inch (FIG. 11-32). Moving the adjustment all the way down will give the minimal walking clearance of about 3/32 inch (FIG. 11-33). The latter is used for lighter fabrics. On this particular machine, the general position that works well for most furniture sewing needs is indicated in FIG. 11-34.

11-30 Rear view of machine: (A) pressure-foot hand release lever; (B) walking-foot height-adjustment mechanism.

11-31 Walking-foot adjustment, shown set for maximum height.

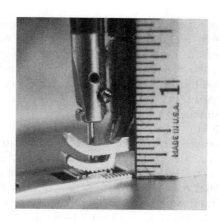

11-32 Walking-foot adjustment shown at maximum height (setting shown in FIG. 11-31).

11-33 Walking foot shown adjusted for minimum height (thumb screw would be set at bottom of groove, FIG. 11-31).

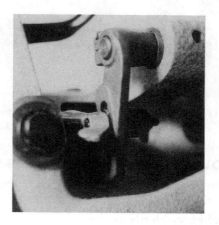

11-34 "Universal" setting for walking-foot height for the majority of upholstery sewing needs.

Pressure-foot pressure adjustment

That impressive assemblage of black steel running lengthwise along the top of the machine is the pressure-foot tension assembly. The two long bands of steel are the adjustable spring units that apply the pressure to the foot to squeeze the fabric tightly together for each stitch. The best tension adjustment is the lightest that will squeeze the materials together tightly without crushing or wrinkling them. The lighter the pressure, the lower the power required to drive the machine and the lower the wear on parts. But, like the pressure-foot height adjustment, rather than change the pressure adjustment for every deviation in sewing thicknesses and fabric "hardness," the adjustment is set to handle the "average" work being done. This adjustment is seldom changed.

If very light, soft fabrics are being sewn most of the time, reduce the pressure. For heavier, tougher materials, increase the pressure. Notice that the spring bars (F, FIG. 11-2) are attached to a vertical shaft at the head of the machine (K), pass beneath a fulcrum (H), and end at the adjustment site (G).

Figure 11-35 shows a better view of the adjusting assembly and the results of moving the adjusting and lock nuts (C). The fulcrum (A) maintains a constant downward pressure on the spring bars (B). Due to the fulcrum action, an upward movement at point C will produce an increase in downward pressure on the needle bar. Downward movement at point C will reduce pressure at the needle bar. Note: Because of the heavy spring tension, it requires some effort to turn the adjusting and lock nuts.

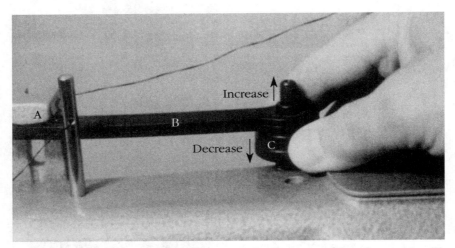

11-35 Pressure-foot pressure adjustment: rotating finger nuts (C) to force spring (B) upward increases foot pressure. Fulcrum (A) creates pivotal pressure on spring.

TROUBLESHOOTING SEWING PROBLEMS

In this section, some of the most common troubles are highlighted.

Stitches suddenly cease

1. Bobbin thread no longer feeding

~Bobbin out of thread: Put in new bobbin.
~Bobbin thread broken: Check for entanglement, knots, and thread
   tensions.

~Bobbin thread intact: Check drive dog. If it is engaged, as illustrated in FIG. 11-36, check for broken needle or timing problem. If it is disengaged, (FIG. 11-37), rotate handwheel while holding the shaft (A) with the other hand to locate latch directly over the groove (FIG. 11-38). Depress the spring-loaded retaining pin, as illustrated in FIG. 11-39. An ice pick, scriber, or jeweler's screwdriver (FIG. 11-39A) can be used to press this pin (B) in toward the large hub (C). This will permit the latch to snap back into the groove ready to drive the pick assembly again. If the pin will not depress, try putting the blade of a straight-slot screwdriver on top of the latch (at point D) and sharply tapping the handle downward with the palm of the hand. This usually will snap it back into the groove.

11-36 Drive dog assembly (for thread picker) in engaged position.

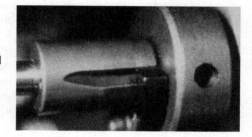

11-37 Drive dog "popped" (out of groove).

11-38 To reset the drive dog, first position the latch over the groove, then depress the spring-loaded pin or snap the dog down.

~Drive dog pops out too easily:

(1) Raise machine head and try rotating the handwheel. With tension off the belt from the motor to the head, the handwheel should turn freely. If it does not, thread is probably caught in the bobbin or picker assembly.

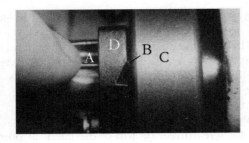

11-39 Dog (D) located over groove ready to reset: (A) scratch awl to depress pin (B) into drive housing (C). If pin is too difficult to depress, a sharp tap where the D is located on dog will snap it back into position.

(2) Check bobbin and picker assemblies for entangled thread. If it looks like thread has been drawn beneath the bobbin case and it cannot be worked out from above, remove the bobbin cap and bobbin (FIGS. 11-9 through 11-11). Remove the bobbin case by

- extracting the three screws in the retaining ring, (FIG. 11-40);
- removing the retaining ring (FIG. 11-41); and
- taking out the bobbin case by lifting straight up on the center post with the fingers (FIG. 11-42;, needle-nose pliers were used only for a clearer view of the action). The entangled thread will immediately be exposed and removable with ease.

11-40 Removing screws (3) to bobbin-case retaining ring. This is necessary when thread gets caught beneath bobbin case.

11-41 Lifting out bobbin-case retaining ring.

If the base of the thread picker seems to be dry of oil or there is an accumulation of lint, it should be cleaned and oiled (the lint pictured in FIG. 11-43 is considered minimal). To accomplish this,

(a) Extract the center screw, FIG. 11-43 (it will be in tight). A handy way to do this task is to release the motor drive belt from the handwheel, then, holding a screwdriver tightly in the slot of the screw, rotate the handwheel so the picker assembly will want to rotate clockwise, the pressure will tend to turn the screw counterclockwise, loosening it. Using the handwheel to apply turning pressure on the screw is a lot easier than trying to get both hands in there to turn the screwdriver.

(b) Position the thread pick as shown in FIG. 11-44 (arrow) by rotating the handwheel.

(c) Lift the assembly out with the fingers at an angle, as indicated in FIG. 11-45. Wiggling the assembly slightly makes removal easier. The drive disk (B in the photo), can be removed easily by lifting straight up. When replacing this assembly, be sure that the drive nib on the disk engages in hole C in the picker assembly.

(3) Increase compression on latch spring. Figure 11-46 shows the back side of the drive-dog assembly with the adjustable tension pin at A and the access entrance to the adjusting screw at B. Figure 11-47 illustrates making the adjustment.

11-42 Removing bobbin case. (Bobbin cap, bobbin, and bobbin-case retaining ring have been removed before this step.)

11-43 Removing mounting screw to thread-picker assembly (necessary only to clean, oil, or replace assembly).

2. Needle thread no longer feeding

~Spool out of thread: Replace with new spool.

~Needle thread broken:

(1) Check for entanglements or improper thread path, (FIGS. 11-1 and 11-2).

(2) Pick may be burred, thus cutting the needle thread. (FIG. 11-44, arrow). Closely examine the point of the pick. Remove it and either replace it with a new one, remove the burr, or sharpen it. The point can be sharpened by using a small fine-grit abrasive stone (slip stone). Do not use a file. The pick is harder than the file.

(3) If the timing is slightly slow, making pick enter and sever the thread, reset timing as indicated earlier.

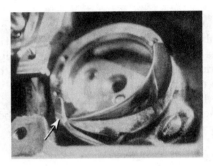

11-44 Proper position of pick to remove assembly. (This is about the only position that will permit removal.)

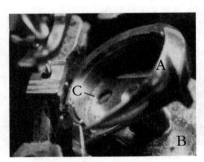

11-45 Remove pick by lifting up at point A to achieve angle as shown: (B) pick-drive plate; (C) indexing hole in pick to receive nib in upper side of drive plate.

11-46 Rear view of drive-dog assembly: (A) spring-loaded latch pin; (B) access to latch-pin adjustment screw.

3. **Chugging sound when sewing.** This is a simple one. Check the overarm take-up mechanism. The thread has probably slipped out of the proper path. Relocate thread to its proper path, as illustrated in FIGS. 11-4 and 11-5, making sure it does not loop around the adjusting screw outside the take-up spring.

4. Sewing tension changes

~Needle thread tension increases:
  (1) Check thread path from spool. It may have slipped down around the base of the spool. This happens too often.
  (2) Check thread path on machine. Reaching over the machine head, shifting of fabric, or other body movements could have altered the normal threading path.
~Bobbin thread tension decreases:
  (1) Check bobbin placement in cap. It may have been put in backward and the thread slipped out from under the spring.
  (2) Check bobbin-thread tension by pulling thread from bobbin cap. If tension on spring is insufficient,
      (a) Try tightening tension adjusting screw. If tension does not increase noticeably
      (b) Remove old spring and replace with a new one, then adjust tension.

11-47 Adjusting tension to drive dog latch.

MAINTENANCE

The best maintenance on any machinery is preventive maintenance. The first rule for long and reliable service is *keep it clean*. Sewing produces significant quantities of lint—fine particles of fabric created by the needle piercing the fabric, thread friction against fabric threads, and pressure and walking-foot motion across the fabric. Figure 11-48 shows a very light distribution of lint beneath the cover plate. The most popular way to

11-48 Bobbin cavity showing a small amount of lint. This would be considered only slightly dirty. *Caution:* It takes relatively little sewing, especially on napped fabrics, to build up lint deposits. Clean frequently.

clean lint is to blow it out with compressed air, then lightly oil the moving parts. The oil will tend to wash migrating lint to the outside. If an air gun is not available, there are commercial aerosol units of clean, dry air. They work well. A fine soft-bristle brush also works well. Figure 11-49 shows a bobbin area recently cleaned with compressed air.

11-49 Bobbin cavity after cleaning with air gun.

The second rule in preventive maintenance, one seldom overdone, is *Lubrication*. A lightweight machine oil should always be used. Every machine has most of the important lubrication spots identified in the operator's manual. If the manual is not available, remember that at any location where a sliding or rotating interface is encountered lubrication will be required. Best results, however, will be obtained by following the manufacturer's instructions from the manual, so keep it handy.

Figure 11-50 shows an operator applying oil from a polyethylene squeeze-bottle oiler to one of the oil "wells" located beneath the head cover plate. That same spot is identified as point A in FIG. 11-51 with five other locations identified by O. In FIG. 11-52, ten oil spots, as seen from the top of the machine, are identified by white or black dashes. Figure 11-53 shows two (marked O) that are reached from the underside of the machine. The two spots marked O in FIG. 11-54 also require frequent oiling. To get to the upper spot requires the removal of both the bobbin cap and case-retainer ring (FIGS. 11-10, 11-11, 11-40, and 11-41). The lower spot in FIG. 11-54 is normally occupied by a small felt pad. It is missing.

All moving parts should be oiled. Frequent light oiling is superior to infrequent heavy oiling. This is *not* a situation where if a little is good, a lot is better.

11-50 A polyethylene squeeze-bottle oiler gives control of oil at all times.

11-51 Six oil locations (A and O, beneath head cover plate): A is position shown being oiled in FIG. 11-50).

11-52 Ten oil locations (indicated by—) from top of machine.

11-53 Oil locations (O) in machine base.

11-54 Two oil locations in bobbin assembly.

SEWING TIPS

Although this book makes no attempt to provide complete sewing instruction, a few tips are included in hopes that they will make this aspect of upholstering more pleasurable and successful (especially if this is your first experience on an industrial machine).

Needle types

There are needles for specialty uses as well as for general-purpose sewing. They vary in size (diameter), length (H3 or H4), and point configuration. Figure 11-55 shows four point configurations. Although some operators may say that any needle will work with every material, some will work better and smoother than others in given materials. Using the needle designed for a particular material will consistently produce superior stitching and reduce the wear and tear on the machine.

11-55 Needle points and their purposes: (A) round, for cloth and vinyl fabrics; (B) wedge, for coarse-weave tweeds; (C & D) diamond, for leathers. Most upholsterers use one needle for all sewing, a round-point, number 137 × 17. Specialty shops or machines used for "special" types of work are about the only ones using the wedge and diamond points.

For most custom shops and general-purpose sewing involving cloth and vinyl fabrics primarily, the round point is preferred. The 17, H3 round point is preferred by most upholsterers for general-purpose sewing. Many upholsterers will use this needle for all work that goes through their shop. Unless a lot of work is to be done on a specific type of material, most upholsterers feel it would be a waste of time to change needles.

Cover plates

Keep cover plates closed when sewing. It is a sad experience to have the thread pick tear into a piece of fabric. Not only is a piece of fabric disfigured, but the machine can be severely wrenched out of time. One machine required adjustments clear back to the main drive shaft near the handwheel because a fast-moving pick rammed into a clump of fabric. Such an unusual adjustment is not only difficult to find but is also very time-consuming. The simple solution is to keep all cover plates closed, especially those over the bobbin, as illustrated in FIG. 11-56.

11-56 View showing machine throat plates. Keep closed for all sewing. Note the ½-inch seam guide etched into the right plate. (Etching was done with vibrating carbide tip.)

Consistent seam allowance

One of the best aids, especially for the novice, is to establish a consistency for *all* seam allowances. A popular allowance is ½ inch. If every seam is given a ½-inch allowance, patterns, cutting, and sewing is greatly simplified. One of the handiest ways to maintain a ½-inch sewing allowance is to etch a line in the cover plates on both sides of the needle. Figure 11-56 shows a machine with the right cover plate etched, the left one unetched. Permanent marking like this can be done by sand blasting, chemical etching, or as in the photo, mechanical engraving. A popular quick-and-dirty alternative to permanent etching is the application of a piece of masking tape, as illustrated in FIG. 11-57. The inside edge of the tape should be ½ inch from the needle point.

Avoiding visible stitches

Most machine sewing is designed to produce invisible stitches. *The top stitch is the exception.* Therefore, a little thought must be given to some work to avoid stitches being seen. For instance, one piece of material may be sewn as many as three times before that sewing task is complete. One such case, and a very popular one (except for very experienced upholstery tailors), is sewing up a welt (FIG. 11-58). Often the welt cord is sewn into the strip, as in this illustration. Then the welt is sewn onto a cushion panel, and finally the boxing and zipper panel are sewn to the welt and cushion panel, making three times that the welt fabric is sewn. If the first stitch is tight against the cord and subsequent stitches not quite as tight (which occurs frequently with beginners), the first stitch is objectionably visible.

11-57 Half-inch seam guide made by using masking tape.

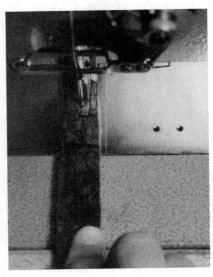

11-58 Sewing a welt. Stitch is *not* tight against cord at this point.

Notice that in FIG. 11-58 the walking foot is not riding tightly against the welt cord, which is to the left of the needle. The welt is sewn loosely at this point. The next sewing, to the cushion panel, will also be done a bit loose (FIG. 11-59). The operator is shown preparing to make a square corner (the purpose of the notch in the welt will be explained below). With the welt securely in place on the cushion panel, the boxing and zipper panel can be sewn to them with greater ease than if all three components were still free to move with respect to each other, especially with napped fabric.

Figure 11-60 shows the boxing being sewn to a cushion panel with attached welt. Notice that the operator is now pushing the welt tightly against the left side of the walking foot. This creates a tight seam and conceals the previous stitches. Notice also that very short runs are being taken. This becomes increasingly important as the nap length increases (napped fabric crawls all over the place unless it is held so it can't).

11-59 Sewing welt to cushion panel. (Stitch is still not tight against cord.)

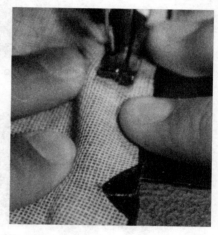

11-60 Sewing boxing to welt and cushion panel. (Now make stitch tight against welt cord.)

Keeping things square & aligned

The universal way to assure alignment of matching panels is using "notches". These may be made in numerous shapes and combinations, the most popular being a single V. In fabrics with a plaid or floral pattern, the pattern can be used for alignment, but it takes time to keep opening the fabric to check. Put in alignment marks or notches. In FIG. 11-61, the operator is clipping off the corner of a boxing folded in half lengthwise. This will create a V notch in the center and is to be made on both sides. A corresponding notch will be made in the front center of both cushion panels (if the cushion is to be a box cushion). These two marks, when aligned, ensure that the centers will line up properly. Other notches are made to permit a periodic check on tensions and spacing, as where the welt or boxing must bend at the corner (FIGS. 11-59 and 11-60, respectively). When the alignment notches have been cut into one side of a boxing, fold the panel in half widthwise and cut corresponding notches in the opposite side (FIG. 11-62). This will ensure that the two cushion panels, for example, will have their corners located at exactly the same spot along the boxing. Figure 11-63 shows a demonstration boxing panel with a double "V" center notch and periodic alignment notches along both sides.

11-61 Cutting side alignment notches in folded boxing.

11-62 Cutting matching alignment notches in opposite sides of boxing.

Eliminating puckers

It is disconcerting to sew up a cushion only to find that the corners are 2 inches out of square and that the main panel or the boxing has a series of unintended puckers and wrinkles. To eliminate such unsavory occurrences, keep an even restraining pressure on all pieces as they are fed into the needle. For example, maybe the pieces are very heavy and cumbersome. The material has to be almost pushed into the needle because the walking action can't pull all that weight and bulk. Two of the best ways to reduce or eliminate completely this frequent malady:

1. Help the material feed through the machine. It must be pulled through, not pushed. While applying the same restraining tension to both (all) panels being sewn, offset that restraining pressure by pulling on the rear of the sewn materials, as shown in FIG. 11-64. This approach has several immediate benefits. It greatly reduces the affect of any differential in

11-63 A demonstration boxing and cushion panel showing double-V center and single-V alignment notches.

feeding that may occur between the table and the foot. It offsets excessive drag created by the restraining tension or the heavy weight of large pieces. And it facilitates more consistent stitch lengths.

2. Take short runs. Sew a few inches at a time; stop the machine; relocate and reset the ingoing material; and sew a few more inches. This takes more time than sewing several feet each time, but it helps to reduce the pucker problem.

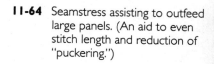

11-64 Seamstress assisting to outfeed large panels. (An aid to even stitch length and reduction of "puckering.")

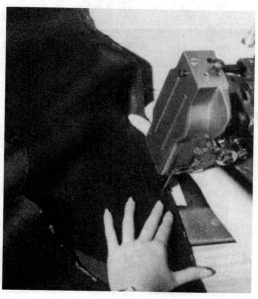

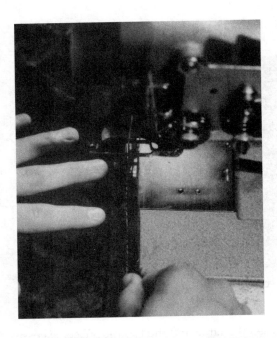

11-65 Sewing fabric with heavy nap. Take short runs. The longer the nap, the shorter the runs.

11-66 Sewing fabric without nap. Longer runs, up to a couple of feet, are possible.

If the fabric has a nap, take very short runs. The longer the nap, the more critical short runs are; the nap seems to make the material wander. The operator in FIG. 11-65 is taking a short run with a crushed velvet. On fabrics with no nap, normally longer runs can be taken, but if puckering seems to be a problem, keep those runs under control at all times. Figure 11-66 shows an operator maintaining positive control of a long run of the ingoing alignment with the right hand while assisting the side-motion and stabilization with the left hand.

Part 2

REUPHOLSTERY TECHNIQUES

Part 2 contains a sequence to follow in reupholstering a piece of furniture. Most overstuffed furniture will proceed from stripping to the application of outside cover panels. Not every piece of furniture will have cushions. Not every piece will have skirts or cambric applications. So not all chapters will be necessary in every case. But each chapter in this part is organized in a how-to order, attempting to illustrate and explain how to get the job done from the beginning to the end.

Each chapter is intended to be complete in itself. For example, chapter 12, Stripping, tells you what to do first, how to keep track of what is done, how to mark the removed pieces, and how to clear the frame of all "gotchas" (staple stubs, wood slivers, protruding nails). Once the unit is stripped, selecting the appropriate chapters will take you through the sequence to the application of the last necessary piece of cover fabric.

It should be recognized at the outset that not every bit of information in every chapter will be applicable to every piece of furniture. However, you should have an awareness of the contents of each chapter so that when questions do arise, the most likely source for answers can be sought.

Part 2

REUPHOLSTERY TECHNIQUES

12
Stripping

Anyone can tear apart a piece of furniture. The trick is to tear it apart so that it can be put together in better condition than it was before the tearing began. And that is what *stripping* really is—not the tearing apart but *the systematic disassembly and removal of unwanted and damaged materials* so that a restoration can take place.

PRINCIPLE OF STRIPPING

The primary principle of *stripping* is to remove only what is absolutely necessary and no more. All too often, those having limited or no experience get carried away in redoing a piece of furniture and end up tearing apart items and pieces that need not be disturbed.

Preparation

A checksheet is highly recommended. The use of some standardized symbols (abbreviations) is also recommended. While "thingy," "do-jigger," or "whachamacallit" might be acceptable when there are no options available, cutting and fitting a panel to the outside arm area when the panel was intended for the inside back is not a gratifying experience, especially when the fabric costs well over $200 a yard and working time is limited—to say nothing about the frustration of trying to match the pattern and color of a discontinued fabric to replace the piece just destroyed. To avoid problems, memorize and consistently use the component identifications listed below.

Component identification

TABLE 12-1 lists component and location abbreviations and their meanings. These symbols have been long accepted in the upholstery trade because of their simplicity and self-descriptive character. It is advisable to commit them to memory so they become like trusted friends, unmistakable and available when needed. A few are listed without their meanings. Try to define those just to convince yourself of just how logical they are. (If not absolutely sure of your accuracy, peek at the answers in the next section).

Table 12-1 Symbols for cover fabric identification.

Component symbols	Location symbols
A = Arm	Bot = Bottom
B = Back	C = Center
Bx = Boxing	F = Front
C = Cushion	I = Inside
D = Deck	L = Left
K = Cambric	O = Outside
P = Panel	R = Right
S = Seat	T = Top
Sk = Skirt	
W = Welt	

Complete mentally or on scratch paper

IA = Inside arm		OA = (?)
IB = (?)		OB = (?)
CBC = Center back cushion		RSC = (?)
RBC = (?)		LCSC = (?)

NOTE: Location symbols precede component symbols.

Procedure

The most expedient stripping procedure is first to take off that panel or part that was put on last and then remember accurately the whole sequence. Good mechanical problem-solving skills and a photographic memory are great assets at this point. But a readily accessible and permanent aid has been devised—the Stripping Checksheet in the appendix. Conscientious use of the checksheet will save many hours of trial-and-error work and frustrating puzzling over "impossible" tasks.

Answers to abbreviations in TABLE 12-1: OA = Outside Arm, IB = Inside Back, OB = Outside Back, RSC = Right Seat Cushion, LCSC = Left-Center Seat Cushion (for units with four seat cushions), RBC = Right Back Cushion.

The general procedure is to start from the bottom and work upward, and from the outside and work inward. The basic sequence is listed in TABLE 12-2 . Although TABLE 12-2 indicates that the cambric (K), SKirt, and Panels may be removed in any order, the cambric *must* be removed before the outside back or outside arms can be started. And often, legs and leg plates must be removed to get to the cambric. Also, in TABLE 12-2 it might seem that stripping begins at the bottom of the unit. It does, generally. Sometimes, however, it might be helpful (if not absolutely necessary) to remove a major portion of the unit before actual stripping begins. Figure 12-1 is a replica of an instruction card relating to a La-Z-Boy recliner chair. Many manufacturers of furniture with removable elements use similar instruction cards, most attached to a frame member on the underside of the unit.

Table 12-2
Stripping and covering order.

Stripping		Covering	
K		D	
SK	Interchangeable	S	
P		IA	Some styles might
OB		IB	be reversed
OA		OA	
IB	Some styles might	OB	
IA	be reversed	P	
S		K	Interchangeable
D		SK	

NOTE: Covering is the reverse order of stripping.

Stripping system

The system goes like this:

1. Look for special disassembly-assembly instructions, especially on recliners and units with very wide arms. Occasionally these tags may get torn or removed by accident, carelessness, or ignorance. If so, it can be extremely difficult to disassemble the unit without destroying it first. A good practice is to examine the unit very thoroughly, looking for levers, nuts, metal brackets, tracks, and clips. Then try to figure out how it is intended to work before forcing something out of shape or into pieces.

2. Prepare a checklist for disassembly and stripping. This will save untold agony when it comes time to recover and reassemble, and is one of the best

TO PLACE BACK IN CHAIR-

Simply lift back to verticle
position at rear of chair.
Align track on back with
mating link on body keeping
both sides in line. With
slight pressure push back
down in tracks until it
stops. Seen from rear of
chair, on the back track
there is a toggle lever—
one each side. With a
household screwdriver engage
toggle lever. Using
screwdriver tip apply firm
downward pressure on toggle
lever. Make sure lever is
in down position for full
lock.
To disassemble, reverse.

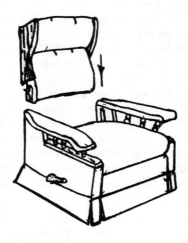

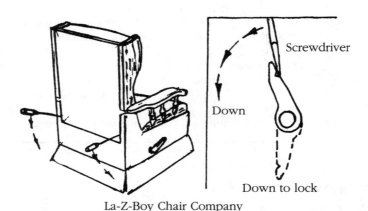

La-Z-Boy Chair Company

12-1 Facsimile of disassembly instructions. Similar instructions should accompany
all units having "special" fastening devices.

problem solvers available. For a sample of one style of checklist, refer to
the appendix. Any checklist can be designed. The important thing is to
ensure a knowledge and memory of what, when, and how it all comes
apart. Only an experienced upholsterer can hope to restore some pieces
if careful notes and even some sketches are not recorded during the
stripping process.

3. Remove hardware pieces (legs, plates, or handles) and removable components (arm, back, or seat assemblies) that are placed on top of and appear to prevent removal of fabric panels. Replace or attach screws, bolts, or special clips to the parts removed or the frame whence they came to avoid an unnecessary puzzle of which ones went where or a prolonged search for the vanished items. Taping screws or bolts near the spot they came out of (¾- or 1-inch masking tape works well) is one way to attach these loose items.

4. Remove and (using the abbreviations of TABLE 12-1) mark with chalk the underside of each piece of old cover material as it comes off. Use the general stripping and covering procedure outlined in TABLE 12-2 as a sequence guide. Keep a careful record of the exact order in which all components (fabric pieces and hardware) are removed. Failure to note the sequence of disassembly may result in acute and perhaps chronic frustration! In this business, taking a shortcut when you are not thoroughly ready for it is like making a 90-degree turn with a car where there is no corner. Something may get bent out of shape.

5. Put old cover panels aside for future reference. *Do not* plan to use these pieces for exact patterns. They are often stretched and distorted. *Do* use them as reminders of fabric pattern and nap orientation, aids to identify where and how the new cover panels are to be placed or sewn, and silent sentinels to ensure that every old piece is replaced with a new one.

6. Save and protect padding and stuffing materials to be reused. These should be put aside where they will not be destroyed, scuffed, or snagged. There is no need to replace old materials in satisfactory condition.

   ~If the stuffing and padding have become filled with dust, even though they may seem resilient, they should be discarded. This is especially true for those having any allergy to household dust. Furniture is constantly bombarded with dust, and even the most conscientious homemaker cannot keep all of the dust particles from being absorbed into the padding materials. *Dust check:* Place padding on a flat surface, floor or table, and slap with the flat palm of the hand. Sudden grayish clouds indicate the presence of excess dust.

   ~Even clean, matted stuffing should be replaced with new material if the unit is intended to have a supple, resilient feel. If a solid, firmer feel is desired, clean matted materials can be covered with a new layer of cotton felt or a layer of Dacron.

7. Check and repair frame and foundation components as needed. Refer to chapter 6.

STEP-BY-STEP SEQUENCE

Probably the most used tool in the stripping process is the patented, double-pronged staple remover (called by some a *Berry picker*). This tool can be used on hardwoods (oak, ash, maple, walnut, particleboard), structural foam (plastic) frame materials, and softer woods (alder, soft maple, fir, poplar).

1. Examine the unit and note on the checksheet features to be remembered, or altered, such as location of seams and welts; sizes and shapes of panels,

bands, and skirts; unusual depressions or lumps that should be changed. Are the seams where they will look best on the new cover? Should there be more or less padding? If so, in what areas? Does the seat insert fit the contour properly and extend the proper distance to the front? These are the kinds of questions to answer and note *before* beginning the stripping process.

2. Remove hardware to get to the last panel of cover installed. The last panel is the one stapled on top of the others.

3. Remove the cover panels in the reverse order in which they were installed. On harder woods such as a particleboard base of a chair, this is accomplished by holding the blade of the picker vertical, placing the point of one prong near one leg of the staple, and applying downward pressure with an up-and-down motion, as illustrated in FIG. 12-2. This forces the point far enough into the fabric and into the hardest of the furniture frame materials to get under the staple for extraction. Watch the position of those hands. One slip of the tool with your hand in the wrong position, and you can cut yourself. Keep your hands away from the front (pointed) end of the tool. Now, while holding the point of the tool under the staple, rotate it to either the right or left, depending on your position and that of the staple and your personal preference.

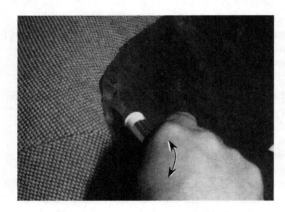

12-2 Getting staple-remover prong under staple in hardwood conditions.

The preferred use of this style of staple remover is to have one leg of the staple locked in the groove of the tool and the tool rotated to the side away from the staple. This procedure tends to pull both legs of the staple out of the wood. This approach also works well on the newer frame materials such as structural plastics (styrene, polyethylene, polypropylene, and occasionally PPO or polyphenylene oxide). On furniture with thick fabrics and softer woods (alder, soft maple, poplar, and gum) the staple remover can be inserted under the staple directly. Although it may take a little more time to get the staple locked, it will save time later.

Frequently in hard materials (oak, ash, walnut cherry, hard maple) and occasionally in softer ones, the staple may break before one or both legs are extracted. The result is "gotchas," those sneaky, pointy little critters that attack the unprotected finger, hand, arm, or clothing.

4. Strip all fasteners on the bottom side of the unit.

5. Proceed to remove the "outermost" panel of material. Figure 12-3 has the operator lifting a folded seam, in this case the seat panel. Note that the underpanel was stapled in place before the outer one was folded over and attached. Figure 12-4 shows the seat panel released and hanging, the pad seat raised to reveal the construction.

6. Remove the fabric from the unit. Notice that with the barrel chair all of the major panels (OA, OB, IB, IA) were sewn together before installing (FIG. 12-5), and are thus removed as one piece. This chair style has only two places on

12-3 A folded seam: sewn at top, stapled on bottom, stretched and folded in between.

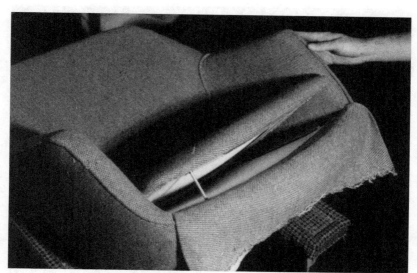

12-4 Seat panel (hanging) with bottom staples removed.

12-5 Sewn slip-over cover (IA, OA, IB, OB panels sewn together). On removal, mark each panel for identification and orientation.

the cover that are not sewn together before fitting to the chair. One is the seat panel, shown hanging loose in FIG. 12-4; the other is the bottom rear joint between the outside arm panel and the outside back, shown in FIG. 12-3. Note also that all individual segments have been labeled. A straight line beneath each symbol is the standard way to indicate the *bottom* of the fabric. Using this notation saves time when it comes to pattern and/or nap orientation.

Most furniture styles have each panel as a separate piece. These are removed one at a time and should be marked as they come off the unit.

A ripping tool (FIG. 12-6) can be used for stripping stapled jobs as well as those done with tacks. Notice that in FIG. 12-6 the tool is placed so the

12-6 Using the ripping tool to strip staples work. Wedge blade under staple and press downward on handle (arrow).

back of the blade is basically flat with the frame surface and under the fabric. Used in this manner, it will tend to lift the fabric and lift out or break off the staple. *Caution:* Great care must be taken when using the ripping tool to avoid gouging the wood excessively.

The ripping tool can also be used for staple cleanup. The corner of the tool is used to wedge under the staple rather than trying to force it out with the flat face. Because most staples are put in so that the back is somewhat parallel to the edge and thus to the grain of the wood, the tool blade will usually be directed against the grain and occasionally toward the edge. If the blade sinks into the wood near the edge of the frame and is tapped with a mallet, the result can be a chunk torn from the edge. Gouges can also be ripped from the flat surface of the wood.

7. Remove the outermost panel. After the cambric is removed, usually the first major panel to be stripped is the outside back. This is accomplished by removing the tacks or staples from the bottom rails, prying the tack strip from the sides by wedging the tool handle downward (as illustrated by the arrow in FIG. 12-7), and removing the tacking strip, which is usually along the top of the back rail (FIG. 12-8). The staple remover is inserted between the outside back panel and the frame and twisted (as indicated by the curved arrow in FIG. 12-8). At the same time, the other hand is pulling outward on the panel. These two actions together make it easy to remove the panel, except in the harder woods.

Once the tack strip has been released a sufficient distance from one end, a faster method can often be employed. Just grab the fabric and tacking strip and pull. The fast method works well with strong fabrics and softer woods but may not work so well with harder woods and more delicate fabrics such as chintz.

8. Check for frame and foundation damage. With the outside back removed, a quick visual check and good notes will identify those internal repairs necessary before new padding or cover is applied. Don't trust the memory only. Notice the "sagging" webbing between the back uprights in FIG. 12-9. In this case, no back padding had to be removed, and the burlap covering the springs was also in good condition. But the webbing had to be

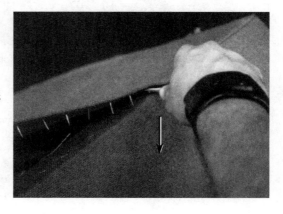

12-7 Prying a tack strip free along sides of OB.

12-8 Removing OB and tacking strip from top back rail.

12-9 Visually inspect frame and foundation for damages and record on check-sheet after each panel is removed.

reinforced or replaced. Reinforcement was sufficient and less time-consuming. To do this, short sections of the burlap along the top back rail needed to be loosened, new vertical webbing strips installed, and the burlap stapled back into position. Chapter 13 shows how this is done. Do not remove more than necessary. Continue with each panel, noting damage and needed repairs as the stripping progresses.

The last cover elements to be removed from most units will be the seat and deck panels. Occasionally the deck may not have to be removed, depending on its condition and the style of the unit. This determination can be made by examining the suspension system and burlap from beneath. If there are no broken or misplaced springs and the burlap is in good condition, leave them alone. Now take a look at the decking and padding from above. If the padding feels smooth and even as you rub your hand across it and the deck fabric is clean, tight, and still strong, leave these alone too.

Figures 12-10 and 12-11 show a chair and set that have a "hard-edge" seat. The hard edge is a strip of wood, ranging from ¼-inch plywood to 1-inch-thick hardwood, depending upon the elevation desired at the front. This strip of wood is attached directly to the front seat rail, padded with one or more layers of cotton, and provides a solid support for cushions.

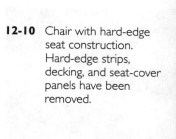

12-10 Chair with hard-edge seat construction. Hard-edge strips, decking, and seat-cover panels have been removed.

In FIG. 12-10 the deck and seat panel as well as the hard-edge strip have been removed. Notice the crease in the cotton about 2 inches back from the front. That indicates where the rear of the hard-edge strip terminated. Figure 12-11 shows the operator removing nails from the ¼-inch plywood hard edge. Half of the padding has been left in the down position to show how it appears immediately after the seat-cover panel is lifted.

In FIG. 12-12 the hard edge has been rolled toward the back to reveal how the cover is tacked to the bottom back of the hard-edge strip (look closely and you can see the staples near the edge of the fabric). Notice that the fabric is stapled about 1¼ inches from the edge. This provides a space for the cotton padding to be tucked under that bottom edge and held in place as the cover is wrapped around (FIG. 12-13) shows more clearly how this was done). This also gives the hard edge a nicely finished appearance.

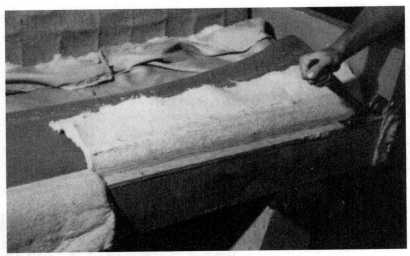

12-11 Removing hard-edge strip from couch.

12-12 Underside of hard-edge strip showing how seat panel is attached.

The last step in stripping this furniture style is removing the seat panel from the hard-edge strip (FIG. 12-14). Note the hand position. If the staple remover should slip, the operator would not encounter one of those stabbing pains.

9. Remove the "gotchas." Now back to the frame. To prevent injury, the gotchas should really be removed after *each* panel is stripped. A closer look (FIG. 12-15) reveals the merciless nature of those stubs. Get them all out. *Do not miss this step!* More agony and irritation result from staple stubs than any other factor in the upholstering business. There are plenty of other gotchas lurking in the furniture without leaving the real demon.

12-13 Padding is wrapped around rear edge of hard-edge strip.

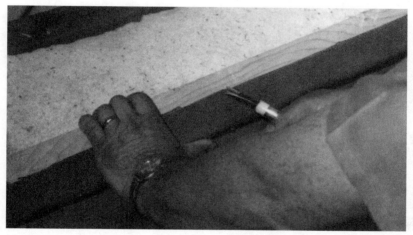

12-14 Removing staples from hard-edge strip. Note safe hand position.

12-15 Close-up of staple stubs. *Warning*: Get them all out or pounded in so they are not protruding in the least! Or pay the sorrowful consequences!

Among the tools best used for this work are the Berry Picker and the dikes. If the stub is long enough to permit locking in the groove, the staple remover works well. However, the diagonal cutter is the best tool for this work. A twisting motion provides good leverage for removing metal remnants with minimal effort. The best cutter for this purpose is one that is slightly dull—not so dull that the cutting edges no longer meet, but dull enough that it takes a little more than normal effort to cut the staples. It is better to remove the staples than to cut them off. The very short stubs left from cutting are real finger-tearers. Even the shortest of them should be hammered into the wood to prevent those painful snags.

Slip-joint pliers can also be used, but they don't seem to grasp the staple pieces as well as the dikes.

10. As the final step, check the frame surfaces for any lurking stubs. If fingers are used, move them very lightly and slowly over the surface. Locate those sharp little protrusions so they can be pounded in or removed. An alternate method is to use the dikes as a detector and slide them lightly across the surface. Any protrusion will show as an obstruction to smooth travel.

13
Foundations

REINFORCING SAGGING WEBBING

Occasionally, the foundation may be in basically good condition, only the webbing is sagging (FIG. 13-1). One remedy is to tighten the old webbing. This can sometimes be done by drawing out the looseness and stapling the excess as illustrated in FIG. 13-2 (use ⁹⁄₁₆-inch staples, nothing shorter). Another way to tighten the old webbing is to remove the top portion of the burlap, detach the top of the webbing, restretch (the webbing plier is probably necessary), and restaple.

In some cases, sagging webbing is reinforced by adding new tight webbing right over the old. This practice is used only when it is obvious no other interior damage needs care. Reinforcing webbing strips are placed along the centerlines of the coil springs. Installed in this way, they provide the most solid support. Reinforcing webbing was improperly installed on the unit in FIG. 13-3. The reinforcing strips catch only the edges of the springs. For maximum benefit, all reinforcing strips should be applied directly in line with the centers of the springs.

Use the following procedure to reinforce old webbing:

1. Place-tack (the new strand to one edge of the frame, in line with the centers of the springs, FIG. 13-2).
2. Fold the flap down and staple securely in place (FIG. 13-3).
3. Stretch the webbing taut and place-tack to the opposite rail (FIG. 13-4; in this case, the stretcher is held in the right hand; the left hand doing the stapling).
4. Cut the webbing about ¾ inch long, fold over the flap, and finish staple as in step 2, above.

Another approach is to attach additional strips to double the effect of those already installed. One such addition is shown in FIG. 13-5. Notice the trifold installation, with the webbing coming from the bottom. This virtually eliminates any tear-out potential. The positioning shown (coming off the back edge of the rail) will give a "softer" back than the midpoint position illustrated in FIG. 13-1.

Attaching the top in this case is simply a matter of pulling the webbing tightly across the inside of the top back rail and stapling. Figure 13-6 shows the place-tacking. Notice also that the burlap has been detached only where the reinforcing webbing is to be pulled through. The webbing should be cut ½ to ¾ inch beyond the back edge, the flap folded over and stapled on top of the main strand to minimize tear-out.

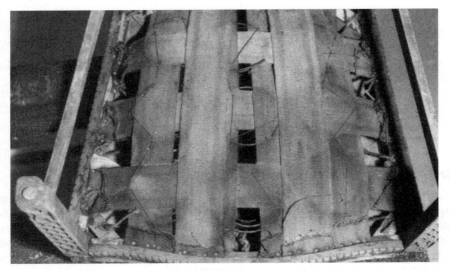

13-1 Improperly placed reinforcing webbing.

13-2 Webbing that has been spot-stapled only.

REATTACHING SPRING EDGE WIRES

A unit with burlap wear as extensive as that in FIG. 13-7 may mean additional problems beneath. Removal of the burlap from the bottom of this chair back reveals detached spring edge wires (FIG. 13-8). This repair can be achieved quickly and efficiently with a heavy-duty pneumatic stapler (FIG. 13-9). Relocate the edge wires, straddle the wire with the head of the staple gun, and pull the trigger; ¾-inch cement-coated staples are recommended.

13-3 Webbing, finish-stapled in preparation for stretching.

13-4 Stretching and spot-stapling second end of webbing prior to cutting and finish-stapling.

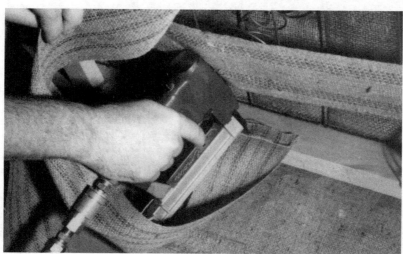

13-5 Stapling first end of reinforcing webbing. Trifold-stapled on top to minimize tear-out.

13-6 Spot-stapling stretched webbing to top rail. Note that burlap has been loosened only where webbing is to be pulled through.

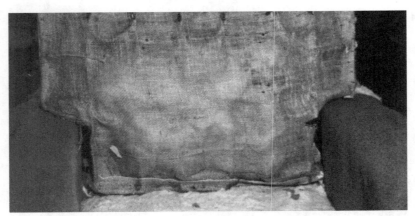

13-7 This burlap should be replaced or covered with new material. *Caution:* Worn burlap may be concealing other foundation problems.

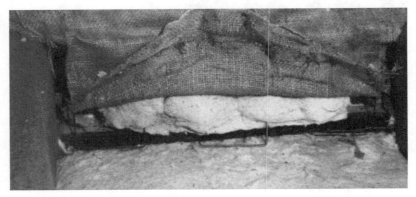

13-8 Back modu-loop spring wires have broken loose.

INSTALLING NEW SINUOUS SPRINGS

Before we touch on installing new springs, let us take a look at one of the fundamental considerations of sinuous-spring suspension systems. The greater the arc in
the spring, the greater the "sit-in" characteristic. That is, there will be more "soft-
give" to the suspension before the solid spring suspension is called into play. Both
the love seat and the wingback chair have rather flat support, although not all
loveseats and wingbacks have flat suspensions. This gives a firm, stable seat. If the
springs had one or two loops added to their span, the seat would feel like you were
sitting into it rather than on it.

The first step to installing a new spring system is planning the layout. Figure
13-10 is a new-frame construction on which the sinuous-spring clips have been located and attached, awaiting installation of the spring sections. (These were attached with the heavy-duty stapler pictured in FIG. 13-9). This sofa could have been
a used sofa that had coil-spring suspension replaced with sinuous springs. The
principle is the same. Determine the most desirable "hardness," then space and arc
the springs according to TABLE 13-1.

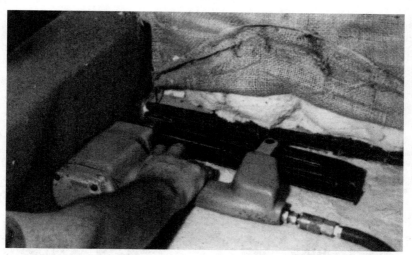

13-9   Reattaching back springs with heavy-duty pneumatic stapler. (Edge wires
can be reattached in same manner.)

Once the layout has been determined and the clips installed (FIG. 13-10), attach
the sinuous springs to the clips at the rear of the unit. Then, with the sinuous-spring
puller, pull them over and into the front clips (FIG. 12-14). A wedge-style spring puller
(made by your author) is being used (FIG. 13-11) to stretch a back spring. (Experienced upholsterers seldom use a puller on back springs; they are relative easy to
stretch.) This wedge puller is especially useful for stretching short, low-arc seat
springs because (1) it requires much less effort by the operator because of its wedging action, (2) it does not mar the frame, and (3) there is no slippage, being lined
with rubber. After all springs have been attached, the heavy-duty pneumatic stapler
is a fast, sure way to secure the top of the clips (FIG. 13-12). Clips can also be installed
using barbed tacks or special nails that can be purchased from supply houses for
this purpose. Figure 13-13 shows the completed seat suspension.

Table 13-1   Sinuous-spring calculations and specifications.

Distance between arms along front seat rail	Number of strands	Center-to-center spacing of clips	Inside arm posts to center of two outside clips	Size of connect links
		Original No-Sag XL		
20"	5	4"	2"	2⅜"
22"	5	4½"	2"	2⅞"
24"	5	5"	2"	3⅜"
48"	10	5"	1½"	3⅜"
52"	11	4¾"	2¼"	3⅛"
72"	15	4¾"	2¾"	3⅛"
78"	17	4⅝"	2"	3"
86"	18	4¾"	2⅝"	3⅛"
		Supr-Loop		
20"	4	4¾"	2⅞"	2¼"
22"	5	4¼"	2½"	1¾"
24"	5	4¾"	2½"	2¼"
48"	10	4¾"	2⅝"	2¼"
52"	11	4¾"	2¼"	2¼"
72"	15	4¾"	2¾"	2¼"
78"	16	4⅞"	2 7/16"	2⅜"
86"	18	4¾"	2⅝"	2⅜"

Space springs on frame by first positioning the two outside strands and then dividing the remaining area in equal parts.

Spacing is to be calculated to centers of clips.

No-Sag Spring Division, Lear Siegler, Inc.

13-10 New couch frame readied for No-Sag springs. Spring clips have been located and attached.

13-11 Using the wedge-style sinuous-spring puller.

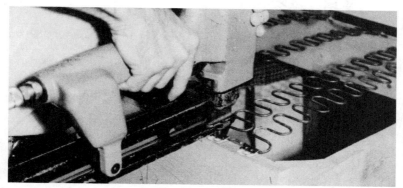

13-12 Locking spring clips with pneumatic stapler.

13-13 Couch seat with sinuous springs installed.

TYING SINUOUS SPRINGS

It is advisable to tie sinuous springs to provide a more consistent base for the burlap and padding. The distance between sinuous ties is a matter of preference. Normally, two ties for seat or back springs will be sufficient. However, if more support is desired, additional ties may be used. A popular tie for this purpose is the "two-and-one tie." Follow this procedure:

1. Attach one end of the twine to the frame at one side, using the double wrap around one tack (FIG. 13-14). Make sure that the short end of the twine loops over the top of the longer segment that will be used for the actual tying.

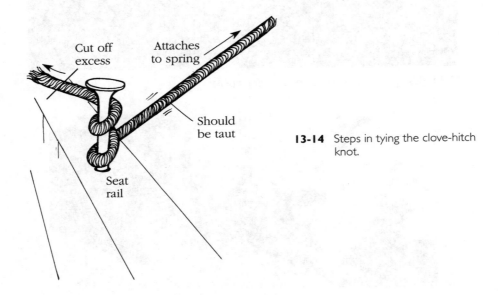

Cut off excess

Attaches to spring

Should be taut

Seat rail

13-14 Steps in tying the clove-hitch knot.

2. Secure it to the nearest loop of the first spring with a clove hitch (FIG. 13-15).

3. From this beginning, go over *two* springs, around *two* loops, back *one* spring, and around *one* loop, as illustrated in FIG. 13-16. (The tie name "two-and-one," is derived from the over-and-around-two, back-and-around-one sequence.) Keep this up all the way across to the other side. The object in this tying technique is to keep the twine snug without distorting the straight-line lie of the spring sections.

4. Tie a clove hitch on the loop of the last spring (nearest the opposite rail from where you began), then secure the twine tightly to the frame just as you started in step 1, above. If a tack is used, use a number 12 or larger. If you use the pneumatic stapler (FIG. 13-17), staple the twine down, loop it back over itself, and staple it again to ensure that there will be no slippage. A double loop can be made around two tacks side by side, as shown in FIG. 13-18. Pay particular attention to placing the loop on the top side of the twine that attaches to the springs. This prevents the tack head from cutting into the portion of the twine that takes all the holding stress.

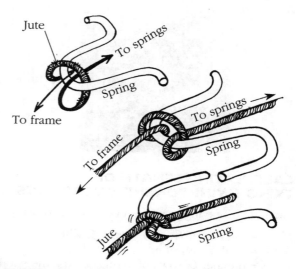

13-15 The two-and-one tie for sinuous springs. (Provides additional support for burlap and padding and equalizes spring action.)

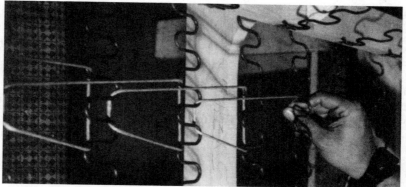

13-16 The double-loop anchoring method for the double-strand tie.

13-17 Attaching end of two-and-one-tie with heavy-duty stapler. (Note clove hitch [encircled] to prevent slippage.)

13-18 How to tie a lock knot.

CALCULATING LENGTH OF TYING TWINE FOR SINUOUS SPRINGS

The length of tying twine can be calculated using the following equation:

$L = (3n - 1)(R/(n + 1) + 3n + 5$, where
L = total length of twine needed
n = number of rows of springs across the width of the unit
R = distance from one rail to the other, taken over the top of the springs

If spring sections are attached to the front and back rails (as illustrated in FIGS. 13-12 through 13-17, R = the measurement over the springs from one side rail to the other side rail. On the other hand, if the springs are mounted from one side to the other side (this isn't normal, but it does happen), then R = the measurement over the springs from the back rail to the front rail.

Example: If a chair back had six rows (n) of sinuous springs (an unusually wide chair) attached to the side posts and the top-to-bottom frame measurement (over the crest of the springs) was 35 inches (R), the twine length would be calculated like this:

$n = 6, R = 35$
$L = (3n - 1) (R/(n + 1) + 3n + 5$
$L = (3 \times 6 - 1) (35/(6 + 1) + 3 \times 6 + 5)$
   $= (18 - 1) (35/7) + 18 + 5$
   $= 17 \times 5 + 23$
   $= 85 + 23 = 108$

If the measurement (R) is taken in centimeters, the length (L) will be in centimeters. If R is taken in inches, L will be in inches.

ANCHORING COIL SPRINGS

Working with coil springs requires more time and effort to tie them than sinuous springs. For that reason, most modern furniture is being manufactured with sinuous rather than coil springs.

If coil springs are to be attached to a wooden base, they will be held in place with tacks (no. 12 or larger), nails, or staples. Frequently, coils will be anchored to a metal strip—the strip and set of springs making one complete row.

Sewing coil springs to a webbing base will work, but it is the old way of doing it and is tedious. Attaching them with the Klinch-it tool is much faster (FIG. 2-31). This is done by straddling the coil wire with the prongs of the clip, having the tool in line with the direction of the wire, as shown in FIG. 13-19, and squeezing the handle while pressing down on the tool. The clip will penetrate the webbing, spread the prongs sideways, and release the clip from the plastic carrier strip in that one operation of the handle. After coil springs are anchored to the base, they must then be tied (across the top of the springs).

The formulas for calculating twine lengths for coil springs are significantly easier than that for sinuous springs. Both formulas below include sufficient twine to tie all the clove hitches, binder and lock knots, anchoring loops, and wraps that will be used.

Double strand tie: Quadruple the frame-to-frame* measurement.

Single strand tie: Double the frame-to-frame* measurement.

13-19 Using a Klinch-it tool to anchor coil springs to jute webbing.

TYING COIL SPRINGS

Tying coil springs has a fourfold purpose: (1) to establish a given height (compression) of the springs, (2) to create a stable base for the burlap and subsequent padding, (3) to hold the tops of the separate springs firmly in place, and (4) to create the flatness or curvature desired. There are a number of ways to tie these springs, but the two described below seem to be the most popular and give the most reliable results.

The double-strand tie

This is the most popular tie for coil springs. The following sequence assumes three rows of coil springs. If there are more than three springs in any row, just repeat the

---

*This measurement is straight-line, from rail to rail, *not* arching over the springs.

knots of steps 2 and 3, 10 and 11 (FIG. 13-27) to the additional springs. The next-to-last spring will incorporate steps 3, and 4, 11 and 12 of FIG. 13-27. All height adjustment and locating of the springs is done with these ties. Normally, the ties are made from back to front for seats and bottom to top for backs, using both strands for each row. If you are tying a back, try placing the unit on its back instead of leaving it upright.

The first strand

Take one piece of twine for the double-strand tie, double it in half, and anchor it to the rear seat rail for seats (or the bottom back rail for backs) with a double loop (FIG. 13-18). From this anchoring point, follow the sequence below. The numbered steps correspond to the knot numbers of FIG. 13-27, which shows the complete tying order.

1. With the first strand, tie a knot (FIG. 13-20) or a clove hitch (FIG. 13-21) around the rear of the third coil from the top of the rearmost spring. Adjust the spring so it is plumb (that is, straight up and down) before you tighten the knot. In tying the lock and binder knots, as well as the clove hitch, the twine is first looped around the spring wire, tensioned as desired, and held from slipping with one finger while the remaining part of the knot is completed and snugged down firmly.

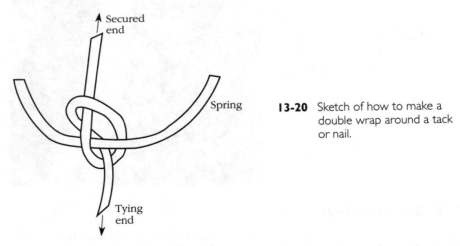

13-20 Sketch of how to make a double wrap around a tack or nail.

2. Tie a binder knot on the opposite side of the top coil of that first spring (FIG. 13-22, right side, shows how a binder is tied). Binders are used frequently because they are fast and easy to tie and prevent any spring movement as long as tension is maintained. A lock knot or a clove hitch will be tied periodically to maintain the nonslipping tension on the binders.

3. Lock tension on the first binder with a clove hitch or lock knot at the back coil of the next (middle) spring, locating the spring in proper position before snugging up the knot (FIG. 13-22, left side).

4. Tie a binder knot at the front, second coil down of the middle spring (FIG. 13-23), right center of the photo.

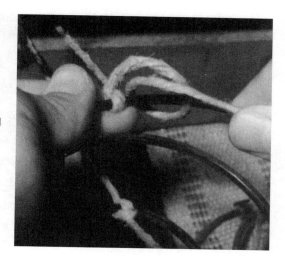

13-21 The clove hitch being tied around a top coil.

13-22 Finished clove hitch (left), binder knot, (right).

5. Go to the front spring and tie another binder at the rear of the third coil from the top (5, FIG. 13-27). Position the spring vertically before snugging this knot down.

6. Move to the front, fourth coil down of the front spring, and secure it with a clove hitch or lock knot. All six ties are shown in FIG. 13-23, with the sequence portrayed in FIG. 13-27. Before securing the clove hitch, check to see that the springs are in line and that those tied on the top strand are at the same height.

7. Anchor this first strand to the front rail by making a double wrap around the first tack (FIG. 13-14) and drive it down tight.

The second strand

8. For the *second* strand, tie a double-lock knot (FIG. 13-24 shows how to tie a double lock) around the first knot tied with the first strand (8, FIG. 13-27).

9. Tie a lock knot or clove hitch on the rear, top coil of the rear spring, pulling it down to the desired height before snugging it down.

13-23 First strand of a double-strand tie in place. The second strand is still lying loosely on the webbing.

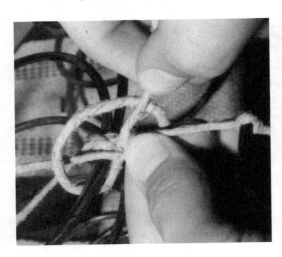

13-24 Tying a double lock knot around a clove hitch.

If the spring foundation is to be flat, the level of this top coil will be the same as the rest of the springs. If the foundation is to be crowned, pull the rear of that spring down to the taper desired and secure in place.

10. Tie a double binder (FIG. 13-25) around the other binder at the front, top of the rear coil (10, FIG. 13-27).

11. Follow the first twine to the middle spring and tie a double binder around the first binder knot (11, FIG. 13-27).

12. Tie a binder around the front of that top coil (FIG. 13-26).

13. At the top rear of the front spring, tie another binder (13, FIG. 13-27).

14. Go down to the front of that same coil and lock around the clove hitch with a double-lock knot (14, FIG. 13-27).

15. Secure the second twine to the frame with a double wrap around a tack or nail.

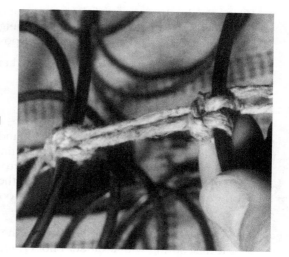

13-25 A double binder tightened around a previous binder.

13-26 A binder knot in the tying process.

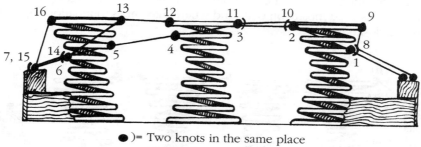

●)= Two knots in the same place

13-27 Complete sequence for the double-strand tie (used on seats, back-to-front tying).

16. Now take the longest of the two twine ends to the top front coil, loop over the coil, pull it down to the desired height (following the same procedure as in step 9, above), and secure that position with a clove hitch.

Do the back-to-front ties on all rows, making each row straight and at the same height.

The single-strand tie

Complete the side ties, using the single-strand tie. (Single-strand ties are used to make all side and diagonal ties; (see "Calculating the Length of Tying Twine for Coil Springs," to obtain the twine lengths). Work from one side to the other. Just be consistent. Figure 13-28 shows the sequence for the single tie. Go to a side rail and anchor one end of the twine with a double loop, leaving a short end of about 10 inches extending from the tack.

1. Begin with the long end and secure the second from the top coil of the nearest spring with a binder knot (1, FIG. 13-28). Pull the coil down to the desired level.

2. Tie a lock knot or clove hitch on the top coil of that first spring (2, FIG. 13-28).

3. Proceed across the top of each spring side, tying binder knots (3, 4, 5, 6, FIG. 13-28).

4. Secure this binder-knot sequence with a clove hitch on the inside second coil of the last spring (7, FIG. 13-28).

5. Tie a binder knot on the outside of the last spring, third coil down (8, FIG. 13-28).

6. Secure to the rail with a double wrap.

7. Loop the remaining length of twine at both sides around the top outside coil, pull it down to the desired height, and anchor it with a clove hitch (10, FIG. 13-28).

Figure 13-29 shows a crowned seat with all the back-to-front, side-to-side, and two longest diagonal ties in place.

Tying springs at four points each (front-to-back and side ties) is commonly used for backs and where light use is expected. It is called the "four-point tie," each spring tied at four points.

When all springs are tied at eight points, it is known as the "eight-point tie." The eight-point tie is used mostly on seats and where maximum foundation support and stability are desired.

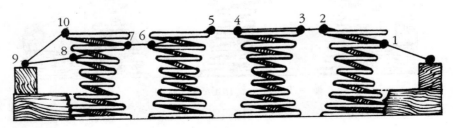

13-28 Sequence for the single-strand tie.

13-29 A crowned seat: all back-to-front and side-to-side ties, and the first two diagonal ties are in place.

Diagonal ties make a smoother foundation for the burlap and add firmness to the spring assembly as a whole. Figure 13-30 shows the first diagonals in place. Figure 13-32 shows all but the corner diagonal ties completed.

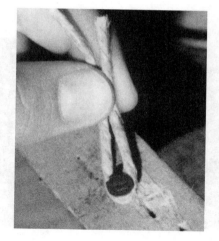

13-30 Diagonal ties are started with a double wrap around one tack.

The diagonal tie

The diagonals are started with a double wrap around a single tack (FIG. 13-30). Notice that only one end of the twine will be used to make all ties. For the diagonals, all ties (probably binder knots) will be made to the top coils only (and the center crossings of the two previous twines if desired). Do not do any contouring with the diagonals. Figure 13-31 illustrates how to get the measurement to put into the single-twine formula for each run—from tack to tack. The tacks are placed so the twine will basically be in a straight line when tying is completed. Figure 13-32 shows the seat with all but the final corner diagonals tied. For crowned seats (or backs), tie as illustrated in FIG. 13-33.

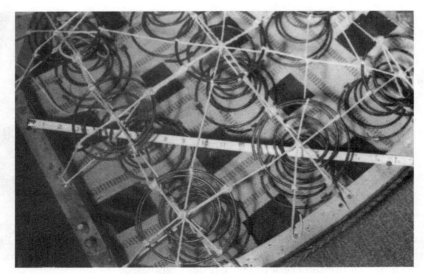

13-31 How to take measurement for diagonal ties.

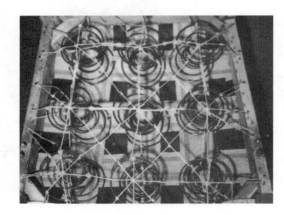

13-32 Completely tied crowned seat.

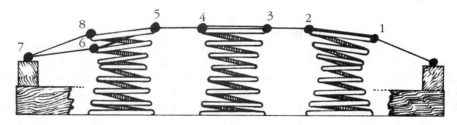

13-33 Sequence for a crowned-seat tying.

REINFORCING COIL SPRINGS

Seats of coil-spring construction (utilizing an edge wire along the front) that have insufficient strength or stiffness can be strengthened by the addition of a homemade V-arc spring. Attach the V arc to the edge wire using Baker clips (FIGS. 2-25 and 2-26). Attach the V arc to the edge wire as shown in FIG. 13-34. Note that the edge wire is located on one side of the center prongs and the spring section is on the other. Figure 13-35 shows the plier almost closed over the Baker clip, binding the two wire sections tightly together.

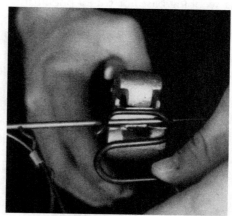

13-34 Baker clip being used to attach a V arc to an edgewire of a coil spring assembly.

The front edge of coil-spring seat suspension units loses some of its strength when people slide forward to stand up; all of their weight is transferred to the front row of springs, compressing them more than the rest of the unit. With increased flexing, increased wear—weakening. This condition is identified by the front edge wire "hanging" below the level of the rest of the seat. Rather than replace the front coils, which are often very difficult to locate, reinforce the coils by adding V-arc sinuous-spring segments.

Determining a V-arc section

When making a V-arc section, recognize that the longer the legs are, the lower the spring support provided. Thus, if you are reinforcing a coil-spring unit that is to be used chiefly by people weighing 110 pounds or less, the legs of the V might well be six inches or more long (assuming a tied height of the coil springs of 5 inches). On the other hand, if the unit is to serve a heavier populace, say, weighing over 200 pounds, each of the legs of the V might be shortened to about 4 inches. If a stiffer front edge is preferred under all circumstances, keep the legs of the V arc short, about 4 inches. The spacing between each sinuous-spring bend is 1 inch, therefore, the lengths of V-arc legs will be in 1-inch increments.

Attaching the V arc

V-arc sections are easily attached to an edge wire using the Baker clip and the Baker plier. If a coil-spring unit needs strengthening and has no front edge wire, one can be attached along the front coils using Baker clips. Then install the V arcs to the edge wire *between* coils. The apex of the V will always be facing toward the spring unit, never outward toward the frame.

With the edge wire and sinuous clip located in the clip properly, squeeze the handles of the Baker plier to crimp the outside prongs of the clip around the wires, as shown in FIG. 13-35. Close the plier completely, and the clip will lock the two wire sections firmly together. Notice that the apex of the V (FIGS. 13-34 and 13-35) is pointing inward. The bottom end of the V arc is attached to the lower edgewire of the coil-spring unit. If there is no bottom edgewire, attach the V arc with a Klinch-it tool (for a webbing base) or with a pneumatic stapler (for a wooden base).

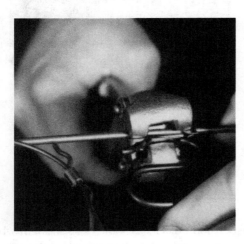

13-35 Baker clip in the crimping process; plier is almost closed for final crimping.

ATTACHING & FITTING BURLAP

One popular method for attaching the burlap is to snug it over the springs, sparingly stapling in place (place-tacking), then fold the extra material (about 1 inch recommended) over the top and finish stapling. This overflap greatly reduces tear-out. The operator in FIG. 13-36 is finishing off the seat burlap (in this case is a surplus piece of durable nylon fabric).

The Y cut

With the spring foundation secured and tied, the burlap base is fitted and attached. The Y cut (FIGS. 13-37, 13-39, and 13-40) is strategic to the fitting process. This cut is used whenever fabric meets a frame member straight-on and is to be finished against the face and pulled around and finished against both sides, or when the fabric is to be stretched in two different directions, as at the arm crest (FIG. 13-37). The top portion will be folded under and pulled outward, around the front of the back upright. The lower portion will be folded under and pulled backward between the back and the arm. The center tab will be folded under to create a finished edge against the face of the arm at the crest. Figures 13-36 and 13-37 show nylon fabric being used as burlap. Notice the way the fabric is cut. If laid flat, it would appear as a Y lying on its side.

13-36 Finishing off a burlap base. (Notice the overlapping that reduces tear-out.)

13-37 The Y cut used to create a finished edge of fabrics (burlap, muslin, cover) against three sides of a frame upright.

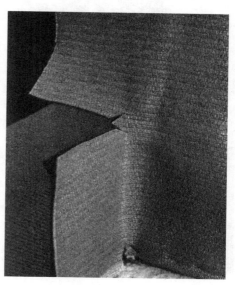

The principle of the Y cut is to cut the center flap, sketched in chalk (FIG. 13-38) so that the cut ends about ⅟₁₆ inch narrower than the chair element the flap is to span. When the flaps are folded under, the corner also folds under a bit, giving a "finished" appearance at each corner (FIG. 13-40, a popular application of the Y cut, fitting around a show-wood arm post).

If a piece of chalk is sharpened (FIG. 13-39), thin-line accurate markings can be sketched with ease. Experienced upholsterers do not take the time to sketch for these cuts. After doing a few hundred units (a dozen for most people), it is almost second nature to know where to make these cuts.

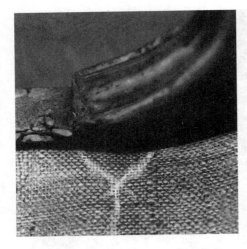

13-38 Marking cut lines for Y cut prior to fitting to an arm post.

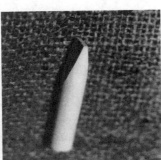

13-39 Chalk sharpened to a wedge-point to reduce width of lines.

13-40 Fitting the Y cut. (Notice the finished edges at front and side of arm post.)

Actually, the Y cut is not really needed when fitting burlap, but if you practice it, fitting the cover will be easier and more accurate. To make this cut (on a seat installation):

1. Stretch and attach the fabric at the center portions of the front and back rails. Leave enough free to fold the fabric from the sides to the face of the frame member that is to be fitted.

2. Mark and/or cut to fit. In FIG. 13-38, the burlap is folded against the inside of an arm post and marked with chalk to show the location and direction of the cuts. The long center cut is to be made straight toward the *center* of

the upright, stopping ¾ to 1 inch from where it is to be fit. The tab in the center is cut so that each leg of the cut will be approximately ¹⁄₃₂ inch short of spanning the width of the frame member on each side ($2 \times \frac{1}{32} = \frac{1}{16}$), and terminate just at the face of the frame member *when the fabric is stretched tight*. For the beginner, it is a good idea quickly to sketch the chalk lines, as illustrated in FIG. 13-38, to show where the cuts are to be made and exactly where they should end.

3. Tuck the center tab under so the folded edge meets the face of the frame member, fold the two flaps under, and stretch tight along both sides of the frame member. Figure 13-40 shows the operator fitting the burlap around the front of the arm post. Notice how the tab smoothly finishes off against the inside edge of the post (right side of the photo).

The diagonal cut

This cut is used when the fabric meets a frame member at a corner and is to be pulled around and finished against just two sides.

1. Stretch and attach the fabric along the center portions of all four sides, leaving enough free at the corners to make a diagonal 45-degree fold across the inside edge of the frame member, as shown in FIG. 13-41. For the novice, sketching the cutting line is recommended, straight from the fabric corner to the edge of the frame member.

13-41 Folding and marking burlap for a diagonal cut.

2. Cut to the corner. For the stout-hearted, skip the sketch and cut diagonally to approximately ¹⁄₁₆ inch beyond where the fabric meets the frame edge. This gives space to stretch the fabric tight to the corner. If the burlap creates a wrinkle at the corner when stretched, make the cut a little deeper.

3. Fold the flaps under and stretch, one at a time, along the side of the frame member to create the finished edges. In FIG. 13-42, the operator is fitting the

13-42 Folding first flap to finish one edge of diagonal cut.

front side of the diagonal cut to an upright with the other flap still left on top of the burlap. The finished diagonal cut is pictured in FIG. 13-43, complete with the burlap overflap, to reduce tear-out, tacked to the seat rail.

For the burlap base, the "finished" edge around upright frame members is not as important as it is for the cover. However, by making the burlap fit neatly, a good habit is formed that will greatly enhance confidence in fitting the final cover, where it counts.

13-43 Diagonal cut finished against two edges of a post.

BASES FOR INSIDE ARMS, OUTSIDE ARMS, & BACKS

Several types of bases are appropriate for the insides and outsides of covered arms, and at times the outsides of backs. For the sake of brevity, this discussion will center on a base for an inside arm. However, the principle remains the same for the outside arm and outside back. This base will most frequently be made from a lightweight cardboard (.020–.030 inch thick). Burlap and webbing are also used on occasion.

The cardboard base

The cardboard base (FIG. 13-44) provides a solid, rigid support for the padding. This is the quickest and most popular technique. The cardboard (.020–.030 inch thick) is sized and cut so that it can be stapled easily to the framing rails and uprights. It must

13-44 Cardboard base for inside arm.

be stapled on all sides to provide a solid base. On some styles, however; there may be a section that cannot be stapled, usually at the rear of the inside arm. In such a case, heavier cardboard (.040–.060 inch thick) is used and stapled only to the top and bottom rails and perhaps the front stump or post.

The burlap base

This type of base provides a full and somewhat resilient support for padding. For those who do not want the possibility of a "cardboard sound" in the unit, this base is recommended. The burlap must be stretched tightly to be effective. It is recommended that the fabric be folded to the outside and stapled in place (FIG. 13-45). Remember: The outward folding minimizes tear-out. To install this type of base,

1. Fold, stretch, and finish-staple the bottom edge in place. Remember to fold a flap *over*; start at the center and stretch it tightly to each side.
2. Stretch fabric to top rail and stay-staple in center.
3. Stretch diagonally up and sideways, stay-stapling from the center along the arm rail. Stay-tacking (stapling) is used at this point because finish-tacking will be done with a flap folded over.
4. Fold burlap flap (approximately 1 inch) over and finish-staple top edge.
5. Snug to center sides and stay-staple in place. The completed inside-arm foundation should look like the one in FIG. 13-45.

The webbing base

This base is strategic where a strong and "forgiving" support is desired or where protection against impact is desired. It is especially suited where knees and elbows may

13-45 Burlap base provides "quiet" support for IA.

be factors with which to contend, as in lounge or living-room furnishings and around which children are likely to be jumping, climbing, or scuffling. For teenagers and for rental units, this is the recommended base. To achieve maximum support, place the webbing pieces almost adjacent to one another.

1. Determine where maximum support and contours occur. Locate and stretch the vertical webbing to provide the most practical support and proper contour.

2. Finish-staple vertical strips, one at a time, at the bottom (fold a flap outward to minimize tear-out). Stretch to the top rail and stay-tack in place.

3. Cut webbing ¾ to 1 inch long, fold over the flap, and finish-staple.

4. Locate horizontal webs to provide the support needed.

5. Attach, one at a time, to the back upright for arms and to one side upright for backs. Weave alternately, over and under through the vertical strips, stretch to the arm post or stump, stay-tack each piece, and cut ¾ to 1 inch long.

6. Finish-staple each piece of webbing. Special consideration must be given to contouring. A close look at FIG. 13-46 will reveal the strategic location of the leftmost vertical strip and the weaving of the topmost horizontal strip under the left vertical strip. Notice how that approach gives the inside of the arm slight convex curvature to approach that of the arm stump.

13-46 Webbing base—quiet and extra strong.

14
Seat & deck

The first panel to be installed on any completely reupholstered piece of integral frame construction is the deck, or seat and deck combination. Only a few of the numerous types of seat assemblies will be *covered.*

THE HARD-EDGE STYLE

The hard-edge seat assembly is usually styled so that the deck and the seat panel are separate entities. The deck is the first panel to be attached after burlap and basic stuffing has been applied to the seat area. Installation is almost identical to burlap, the main difference being that folding a flap over or under to avoid tear-out is not done. It is tacked straight to the frame.

DECK INSTALLATION

Figure 14-1 illustrates the beginning of deck installation. After stapling the front edge to within a couple of inches of the arm post or stump, stretch to the rear seat rail and staple the center (FIG. 14-2). Fitting then proceeds as for burlap installation; including the Y and diagonal cuts. Figure 14-3 shows the decking marked for making the diagonal cut for the back post on a sofa. Figure 14-4 shows the first tuck folded, with the second to be fitted. The decking should fit snugly against any uprights around which it is to be finished (FIG. 14-5).

To fit around the center upright of a couch having a triangular brace (best shown in FIG. 14-9) requires a slight modification to the diagonal cut. In FIG. 14-6, a mark has been projected from the center of the rear upright, then angles over to the front of the brace (FIG. 14-7). This creates a flap for both sides that will be folded under to give as straight a pull on the decking as can be arranged. Each side should be fastened snug to the upright, as shown in FIGS. 14-8 and 14-9.

INSTALLING THE SEAT PANEL

With the deck in place (FIG. 14-10) the hard-edge seat panel is ready to be installed. For specifics on how the hardedge is usually constructed, refer to chapter 12 (Figs. 12-19 through 12-22). With the seat panel attached to the wider plywood strip, nailed in place, and the second or elevator strip also in place, pad the hard edge as desired. In all padding, the outer layer should be completely smooth; it must be the layer that covers the full width and length of the area in one piece if possible. If padding wider than normal is required, splice the cotton so that no ridges will be perceivable, as illustrated in chapter 10 (Figs. 10-16 and 10-17).

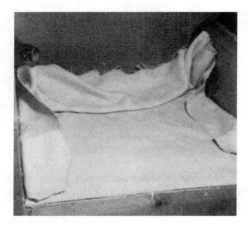

14-1 Attach decking along front edge. Start from center, stretch and tack toward each side.

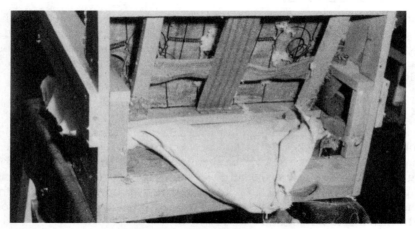

14-2 Stretch and attach center at top of back seat rail.

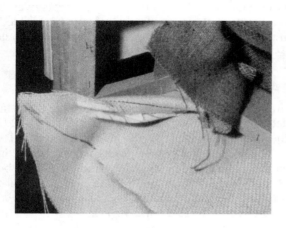

14-3 Back of decking attached almost to corner post, diagonal line marked for cut.

14-4 Fold flap under and form tight to upright.

14-5 View of properly fitted decking at right rear post.

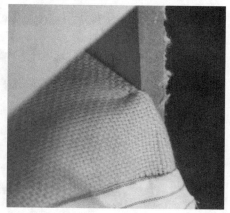

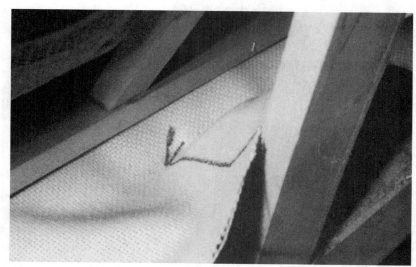

14-6 Marking for modified cut around braced, center upright of a sofa.

14-7 Appearance of modified cut prior to fitting.

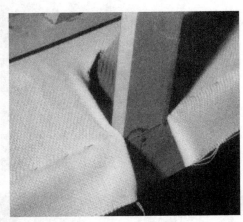

14-8 Properly fitted decking around center upright.

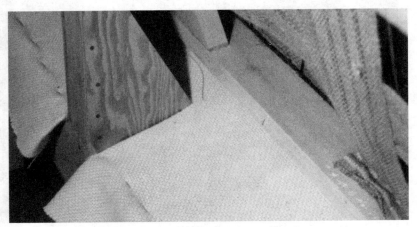

14-9 Finished edge of decking alongside of plywood brace.

14-10 Finished decking on a chair hard-edge construction.

Figure 14-11 shows a narrow layer of cotton felt that will just cover the hard edge and rail edge. At least one more layer of cotton is necessary, and if it is to be the last layer, it should be wide enough to cover the entire front of the rail, as depicted in FIG. 14-13. In this installation, three layers of cotton were desired, so the second or intermediate layer is not full length (FIG. 14-12). The outside and final layer is tucked under the cover at the rear of the hard-edge strip and laid extending toward the rear of the chair (FIG. 14-13). The other two layers (in this case) are then placed on top of it at the front edge, as *might be discernible* in FIG. 14-13 (it is very difficult to see the

14-11 A narrow band of cotton (first layer) for padding hard edge.

14-12 Second layer of cotton, wider than first (for seat).

14-13 Third and final layer of cotton to pad hard edge. (All three layers have been laid back onto deck, revealing hard-edge strip.)

separations of cotton layers). The seat panel with the cotton is rolled forward and pulled over the front rail to be stapled in the center of the bottom. Figure 14-14 shows a completed chair with the hard-edge construction and recessed arms. This style will require a Tcushion. The unit pictured in FIGS. 14-11 through 14-13 has flush arms and will accommodate a standard rectangular cushion.

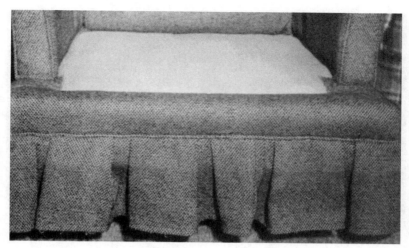

14-14 View of a finished hard edge with sewn cushion retainer groove.

CUSHION RETAINING GROOVE

Regardless of the style of arms and whether the unit has a hard- or soft-edge seat, all units with removable seat cushions will require some means of retaining the cushion in place—the *cushion retaining groove*. This is actually a recessed groove just behind an elevation at the front of the seat. One way to achieve the elevation is using wood strips and padding, as indicated above. Another approach is to establish the cushion retaining groove itself. This can be done by sewing the deck and seat, along the seam, to the burlap and spring assembly, (illustrated in FIG. 14-15). Sew all the way through any padding, the burlap, and wherever possible around suspension spring wires. Make the stitches tight and straight. A double-pointed straight needle is usually used to make a running stitch along the seam. A finished groove is shown in FIG. 14-16.

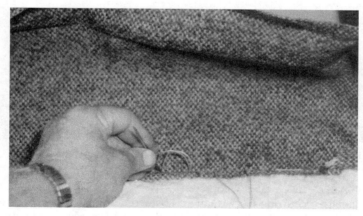

14-15 Sewing seam of decking and seat panels (for cushion retaining groove) through padding and around spring suspension wherever possible.

14-16 A closer view of the sewn cushion retainer groove for one style of hard-edge construction.

Figures 14-17 and 14-18 show another style of a sewn cushion retaining groove. Yet another popular means of establishing the retaining groove, one that is gaining popularity because it reduces hand sewing time, is the use of covered buttons. Figure 14-19 shows the executive chair being prepared using the button method. The finished products are shown side by side in FIG. 14-20. One has the buttoned groove; the other the sewn groove. What difference does it make how the cushion retaining groove is made? Personal preference. No one can distinguish the difference on a finished unit without lifting the loose T cushion. They both work well in retaining the cushion in the seat.

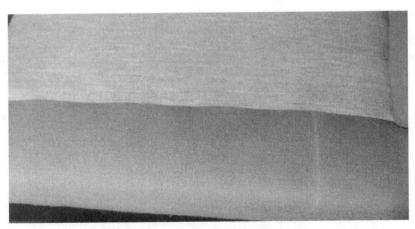

14-17 View of a modern styling for sewn cushion retainer groove.

Occasionally, a deck may become excessively soiled or otherwise damaged with no other damage occurring to permanent cover panels. In such cases, replacing the decking is the only job to be done. With little time and disturbing only the outside arm and outside back panels, a new deck can be installed by using the blind stitch along the deck-seat seam.

14-18 Executive chair utilizing sewn cushion retainer groove.

14-19 Using covered buttons to establish a cushion retaining groove.

14-20 Two executive chairs, one with sewn, one with buttoned retainer groove. (There is a slight difference in the cushion and arm styling.)

THE BLIND STITCH

The *blind stitch* is used wherever machine sewing is impractical or impossible, on-the-unit repairs are to be made, or closing of tuck and pleat folds *on* the unit is desired. Attaching a new deck to the seat panel is shown in FIGs. 14-21 through 14-25. Notice the skewers (FIG. 14-25) holding the folded front edge of the deck panel in place. This step is a great timesaver and aids in ensuring a straight seam.

14-21 The blind stitch to attach a new decking to existing seat panel. (Notice the skewer holding the folded decking in place.)

14-22 Blind stitches are made parallel and close to the seam. (Notice the tip of the needle just emerging.)

The principle of the blind stitch is to make square (90-degree) alternating stitches across the seam that permit tightening the folded edges of the two pieces of fabric together. This stitch is always done with a curved needle. Any deviation from the 90-degree stitching will result in undesirable puckers and gathers. (Additional instructions and illustrations on the blind stitch are given in chapter 9).

1. Start the stitch in a location where the knot will not be visible when the unit is finished and in normal use: (a) near the seam but attached to the old decking that will be covered by the new deck, (b) from the back side at the seam of the new deck fold (illustrated in FIG. 14-21), (c) from the underside of the seat panel itself, or (d) toward the outside arm rail where it will never be seen without taking off the outside arm panel.

2. Exit near the seam to be created (FIG. 14-21).

3. Make the first stitch by moving the thread straight across the seam to the adjoining panel; enter that side exactly beneath the thread (FIG. 14-22).

4. Rotate the needle parallel to the seam and exit fabric the desired stitch length away (¼- to ¾-inch, depending on the type of fabric involved). Use the longest stitch that will give a nice seam (without wrinkles or bunching).

5. Cross the thread, holding the stitches taut, over to first fabric, again at 90 degrees to the seam, and enter the needle exactly beneath the thread (FIG. 14-23).

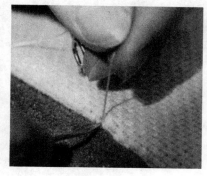

14-23 Blind stitches are made square or perpendicular to the line of the seam.

6. Rotate the needle parallel to the seam and exit fabric at desired stitch length (FIG. 14-24). Continue to end of sewing.

7. Secure the last stitch somewhere beneath or inside of the seam so the knot will be secure and hidden from view. Figure 14-25 shows a couch with new decking partially sewn in place (at both ends) to the old seat panel.

THE SLIP SEAT

Covering a *slip seat* is a straightforward process with just a few tricks that make the process easier. The seat illustrated in FIGS. 14-26 through 14-30 has sewn front corners and tucked rear corners.

1. Slip the cover over the frame and padding. Then open a portion of the corner seam to the point where it just meets the bottom edge of the frame.

2. Snug the side panel into the frame, tightening the fabric on an angle toward the front, as indicated in FIG. 14-26. Make adjustments to ensure that seams are right at the corners of the frame.

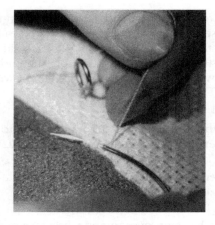

14-24 View showing constant parallelism for proper blind stitch.

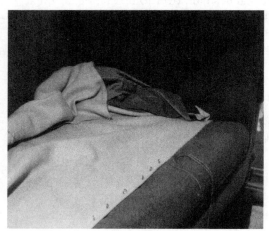

14-25 Blind stitching new decking to good cover requires loosening only the OA and OB panels.

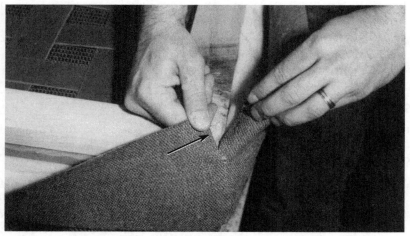

14-26 Opening and fitting the side of a front corner seam for a slip seat.

3. Snug the front portion of the opened seam so that the folded edge will lie at a slight inward angle, taking up the excess material along the front edge. Figure 14-27 shows the side panel stapled in place and the operator beginning the front fold. Fold additional material under if necessary to permit tightening the front panel sideways without letting the fold extend beyond the edge.

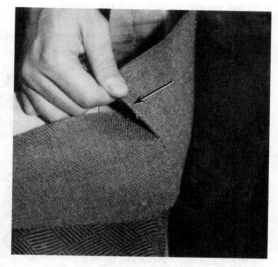

14-27 Fitting (pull in direction of arrow) front corner flap over stapled side.

4. Stretch and tack the front edge in place, then go to the back. Stretch and tack the center in place.

5. Stretch and tack both sides, working from the centers toward the corners, leaving about 4 inches free from each corner, as shown to the right and left of FIG. 14-28.

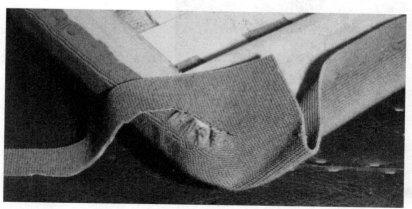

14-28 Fitting a rounded corner to slip seat.

6. Fit each of the rear corners by starting at the side, taking small tucks and stapling each as progress is made around the corner to the back side. Continue around the rear corner to a point where by taking up the excess material a fold will be made that finishes a neat corner (FIG. 14-28, center).

7. With the exact location of the fold identified (FIG. 14-29), cut out the excess material to allow a ½-inch tuck (FIG. 14-30).

8. Finish stapling remaining material down to the seat and trimming off the excess, as shown along the left side of FIG. 14-28.

14-29 Determining where to make fold for a tucked front edge of a slip seat.

14-30 Cutting out excess fabric to make low-profile tuck.

15
Inside arms & backs

One of the major hazards for the novice in re-covering a unit is the tendency to think that time can be saved by making the contour, size, and fitting cuts (diagonal cut, Y cut) in a panel before putting it on the unit. Slight differences in padding and in stretching tensions can make significant changes in exact locations of cuts. Sometimes, even anticipating contours and rough-cutting them before fitting can result in erroneous contours. The best practice is to do the fitting to the unit.

Never cut panels to contour *before* fitting to unit.

FITTING THE INSIDE ARM (IA) PANEL

Figure 15-1 shows a couch arm that has been wrapped with a ⅜-inch thick layer of Omalon carpet padding. Over the top of this, after it is cut and fit, will be a 1-inch layer of 35 IFD, HR foam and a layer of cotton. Then begins the fitting of the IA panel. If you haven't already planned the orientation and layout of every panel for your unit, review chapter 4 before doing any cutting. Place the *rectangular* IA panel over the padded arm. For ease and speed in fitting, use the *four-point stay procedure:*

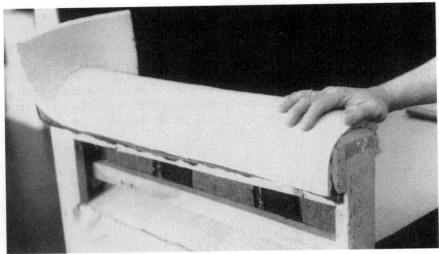

15-1 A rounded couch arm with first layer, ½-inch Omalon carpet padding. This makes a firm nonbottoming under padding for an arm.

1. Stay-tack the front top in place, as shown in FIG. 15-2. This is the *first* of the four-point stays. Smooth the cover to the rear and fold over at the very front edge of the bottom back rail. Mark and make the Y cuts that are essential to fit the cover to the bottom, back rail, and along the crown of the arm (FIG. 15-2).

2. Roll the top portion of the IA not quite half-way under the back extension, as shown in FIG. 15-3. The extension is usually cut to leave a ½- to ¾-inch gap above the arm itself. Make the Y cut so the legs end about ³⁄₁₆ inch short of the front of the solid upright at the rear of the extension. This will result in a total of four "strips" that are to be pulled and stapled to the back upright.

3. Pull the top outside portion of the arm panel back, smoothing with one hand (FIG. 15-4) as you do so to get it quite snug. Stay- tack to the back side of the rear upright. This creates the *second* four-point stay.

4. Go to the inside bottom of the IA to fit it around the bottom back rail. Notice the tab ready to tuck under at the front edge of the rail (FIG. 15-5). Slide the remaining strips through the openings: one beneath the bottom back rail (the one that is shown still laying on the deck in FIG. 15-5). Tuck a small portion of the rear of the IA panel between the deck and the arm, as the operator is doing in FIG. 15-6—just enough to hold it in place temporarily.

15-2 IA panel stay-tacked at front, Y cut to fit at crown and lower back rail.

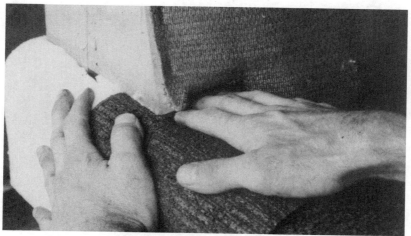

15-3 Rolling top of IA halfway under back extension to locate Y cut.

5. Fold the bottom front portion of the IA panel up over itself to locate the Y cut for the inside of the arm stump or post. *Caution:* Allow enough material so that the legs will not quite tighten against the stump when the panel is pulled tight. If the seat is a soft seat (front of the seat depresses [spring suspension instead of hard]) make sure that the legs will still be concealed when the seat is depressed to its bottommost position.

6. Finish tucking the bottom inside flap of the IA panel between the deck and the lower arm rail. When properly done, it will look like that shown in FIG. 15-7.

7. With all the cuts made and the flaps tucked in their proper direction, smooth the fabric so that the pattern or weave will be oriented vertically and not at an angle. Figure 15-8 shows the operator pulling the fabric

15-4 Smoothing IA panel toward back for the second of the four-point stay-tacks.

15-5 Tab at bottom back rail ready to tuck under.

diagonally downward and forward with one hand while the other is smoothing the fabric forward. This smoothing action greatly reduces the pulling effort needed to get a snug fit. Notice that although the weave seems to be angling forward at the bottom, the pattern will line up when the center portion is fitted. A little care at this point pays high dividends in the end. Figure 15-9, with the pronounced plaid pattern, vividly illustrates the straight vertical orientation so essential to quality work. Notice that the

15-6 Tucking IA panel at decking in preparation to locate Y cut for the front stump.

15-7 Inside arm panel tucked in, ready for third and fourth stay-tacks.

decking is the same fabric as the cover. Seldom is any effort made to match the pattern of a deck to the arm, back, or seat panels.

8. Stay-tack the bottom inside front in place. This is the *third* four-point stay (see FIG. 15-10 for positioning).

9. Complete the last four-point stay by tacking the outside front (also indicated in FIG. 15-10). Notice how the downward stretch on both inside and outside has almost formed the rounding for the top of the arm. With the four-point stay in place, final fitting and tacking is greatly simplified.

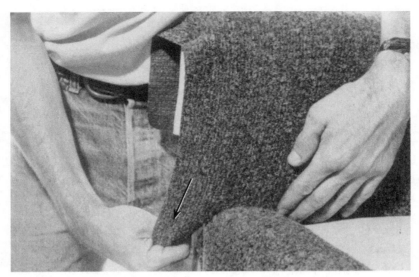

15-8 Smoothing and fitting inside front for third of four-point stay.

15-9 Vertical orientation of IA pattern or weave is essential to quality work.

10. Go to the rear of the unit and snug the bottom strip, ensuring that the tab seats against the bottom back rail. Staple in place (FIG. 15-11). Next, tighten and staple the middle strip in place, as indicated in FIG. 15-12. Complete the inside portion of the rear IA by tacking the third strip (FIG. 15-13).

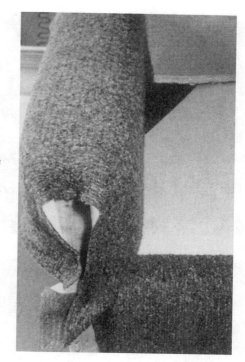

15-10 Fourth of four-point stay in place. Notice how the front of the arm is practically fitted with just the four stay-tacks in position.

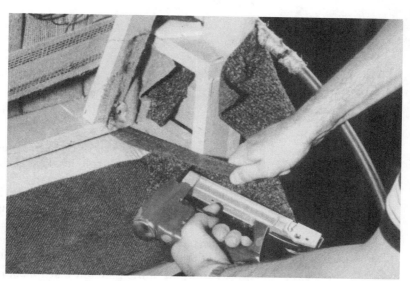

15-11 Fitting and stapling the bottom, rear strip of IA panel.

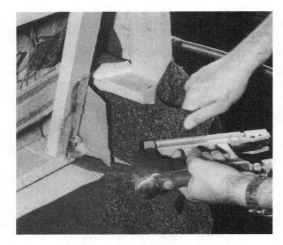

15-12 Attaching middle strip of IA panel.

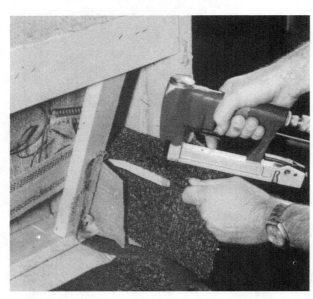

15-13 Snugging and attaching top strip of IA panel.

11. Remove the second stay-tack at the outside rear of the arm if necessary, tighten the outside bottom of the panel, and staple in place at the rear of the back post (FIG. 15-14). Notice the staple holding the panel just beneath and to the left of the staple gun. This staple *must* be below the level at which the outside arm panel is to be attached to conceal the staple when the OA panel is fitted.

12. Finish stapling the rear of the panel in place. By folding any excess underneath, all wrinkles and looseness in the fabric can be removed.

13. Snug and staple the outside portion of the panel in place, making sure that all staples will be below the line where the top of the outside arm is to attach.

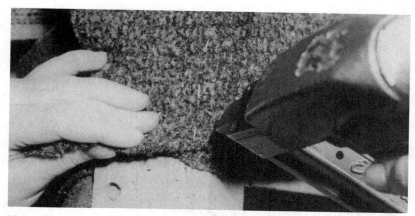

15-14 Tucking and stapling rear outside portion of IA.

14. Return to the front and form and staple both sides to the stump. Notice that stapling of the inside portion proceeds from the bottom toward the top (FIG. 15-15) but does not go all the way. Leave about 3 inches from the top to finish last. The top (or corners) is the *last* portion to fit. When the precise position of this last fold has been identified, lift the folded material to staple the bottom edge in place (FIG. 15-16). Return the fabric as shown in FIG. 15-15 and staple.

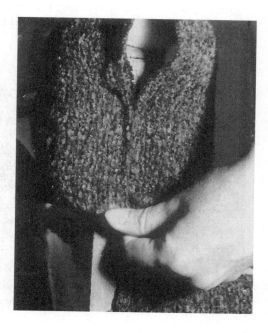

15-15 Final fitting of outside portion of IA panel to front arm stump.

15-16 Making the undertuck to hold front outside portion of IA in place.

15. Now go to the top and start the problem-solving process to determine exactly where the tucks are to be made. This is done by experimenting to see what looks good to *you*. There is no absolute. Whatever is done, be sure that all tucks will result in the folds facing downward so they do not create dust and crumb catchers. Figure 15-17 shows the top being formed. In this case, there are only two tucks to form the top. This gives a mildly squared or flat-top look to the arm.

15-17 Making the final tucks to form the top portion of the IA.

Figure 15-18 shows an arm that will have a more rounded top. The operator is adjusting the inside front of the panel to make the *second* stay-tack. (The *first* stay on *this* arm was made near the top of the outside front, just beneath the foam—(FIG. 15-18). In this case, the operator is smoothing the fabric up the outside, around and over the top, and is pulling straight down to assure alignment of the stripes. Angled stripes at this point will destroy the final appearance. The same down-and-in diagonal snugging into the seat panel results in a smooth, tight appearance (FIG. 15-19). Smoothing toward the rear (FIG. 15-20) for the *third* stay tack is to ensure a tight fit and a straight pattern alignment.

15-18 Assuring alignment of pattern for a rounded inside arm.

A rolled front edge of the arm stump is being finished in preparation for a narrow panel that will be used to finish the front of the arm stump (FIG. 15-21). Notice also that the top is still the last portion to be finished. Final tacking of the outside bottom of the arm is shown in FIG. 15-22. To finish off the arm front with a rounded top, multiple tucks are taken, close together, and stapled inline or at a slight angle, as illustrated in FIG. 15-23.

INSIDE BACK INSTALLATION

The couch shown in FIG. 15-24 has been reupholstered from the frame out with all new materials. It is of the hard-edge design and is now ready for the installation of the inside back panel. A 9-inch-wide channel construction, made with 1-inch foam, was chosen.

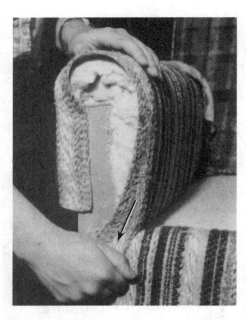

15-19 Angular pull to form and tighten inside bottom portion of IA panel.

15-20 Smoothing for the final adjustment for tacking outside rear portion of IA panel.

When installing an inside back panel, a handy trick of the trade is to lay the unit on its back so the panel will lay naturally on the padding (FIG. 15-25). (The centers of the top and bottom back rails should have been marked before putting the unit on its back.) The center of the back panel should also be marked (or at least identified) to ensure alignment of the panel to the frame. With the above details taken care of, proceed in the following manner:

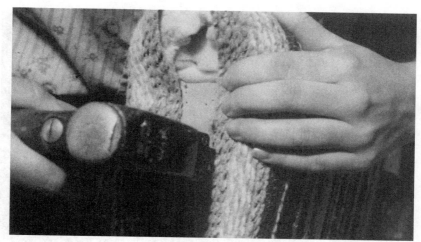

15-21 Forming rolled edge at front of arm stump.

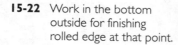

15-22 Work in the bottom outside for finishing rolled edge at that point.

1. Lay the panel on its back, as illustrated in FIG. 15-26, so the foam will just reach the back edge of the top rail. Make sure that the center of the top of the panel will align with the center mark on the frame. After making sure that there is sufficient material to tack at the top, tuck the center portion of the bottom in, as illustrated in FIG. 15-26. Notice the chalk mark on the decking. That is the center mark for the bottom, to help in alignment.

15-23 Rounded arm is formed with multiple tucks that are stapled diagonally.

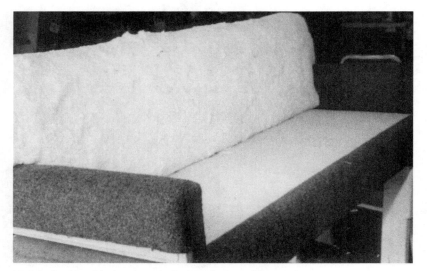

15-24 Back padded and ready for installation of IB panel.

2. Tuck the bottom in, progressing from the center toward the ends, leaving a foot or so free for fitting to the arms. Cut the foam approximately 1 inch oversize so that it will press against all parts of the arm, as shown in FIG. 15-27. Note that the backing is cut out along with the foam. (The piece lying on top of the back panel in FIG. 15-27 is the cutout.)

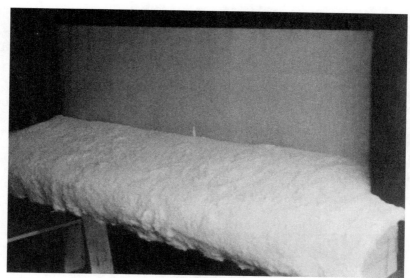

15-25 To make fitting of IB easier, lay unit on its back.

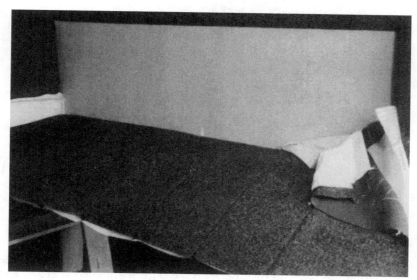

15-26 Laying and tucking a channeled back in place. (Notice the chalk center mark on decking. It is important.)

3. Make the Y cuts as illustrated in FIG. 15-28. The two cuts shown are to provide a pulling strip (A) along the inside of the arm, (B, marked IB with chalk) to cover the back above the arm around to the back post, and (C) a pulling strip that goes beneath the bottom back rail. The top cut is located so that it would lead directly to the arm crown if it were projected all the

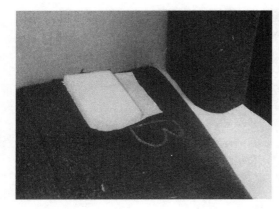

15-27 Corner of padding and backing fabric cut out for arm.

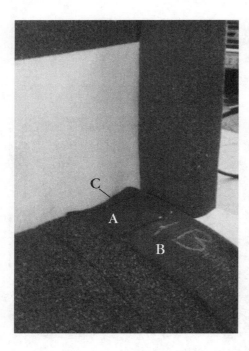

15-28 Y cuts made in IB panel to fit inside arm at crown and bottom back rail. Flap (C) goes under the rail, (A) along inside of arm and (B) around outside to back post.

way. Notice that the cut *does not go all the way to the arm.* This is essential to permit the fabric to tuck beneath the back foam and be totally concealed from view, even when the back is depressed.

The bottom cut is made so that the tab will meet the front of the bottom, back rail when pulled taut. Tuck the cover in and locate where it meets the rail; then pull it out again to make the cut.

4. Tuck the pulling strips (A and C, FIG. 15-28) through their appropriate spots at both ends to continue fitting the back. It will appear something like FIG. 15-29. Then place the unit on its base (the legs have probably been removed) to attach the top.

15-29 Inside back is fitted to arm, bottom tucked in, and is ready to fit along top rail.

5. First stay-tack the IB center to the frame center (identified by the C with the letter *L* going through it, FIG 15-30). In this case a channel seam was center. Marks have been made on the frame for each of the other channels (9 inches). Stay-tack each channel seam in place, as shown in FIG. 15-30.

6. Start in the center of each channel and forming the top to a consistent height, staple to within about 1 inch of each stay-tack (FIG. 15-31). Repeat this process for all channels.

15-30 Stay-tacking channel seams to premarked top back rail.

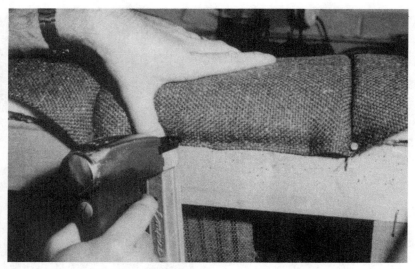

15-31 Level and staple each channel, working from the center toward the seams.

7. Now go back and remove all of the stays. The channel back will appear as in FIG. 15-32.

8. Fold the excess fabric from one side so the outer edge of the fold will lie on the mark (FIG. 15-33; the mark is barely visible beneath the dangling thread).

15-32 Close-up of how seam looks when ready to be fitted.

15-33 Folding excess fabric from one side over to meet at seam mark (not visible).

9. Fold the opposing side in like manner (FIG. 15-34) and staple in place. The finished channel will appear in FIG. 15-35.

10. Go to the extreme sides and trim off the outer edge of the foam at a 45-degree angle, as illustrated in FIG. 15-36. This will give a feathering effect and a smooth rounding as the cover is pulled around and tacked to the back post. Follow fitting procedures much the same way as for the IA panel.

15-34 Properly folded channel seam, ready for stapling.

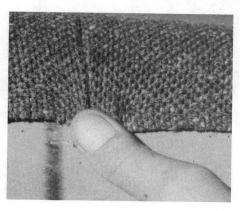

15-35 Completed channel seam at top back rail. (Work from center toward the ends.)

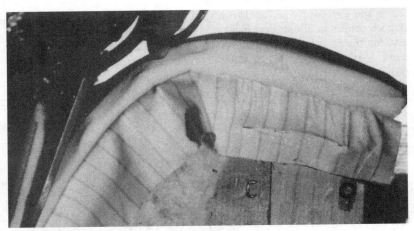

15-36 Trimming outer edge of foam at a 45-degree angle to make a smooth rounding.

16
Outside arms & backs

The outside panels are easy to work on—also exciting, like putting the capstone on a monumental work.

OUTSIDE ARMS

There are dozens of shapes for outside arm (OA) panels, but the same principles hold true for all. In fastening outside panels, the majority of work will be blind-tacked. That is, the panels are to be attached, but the means of attachment will not be visible. The three popular blind-tacking methods are accomplished with (1) tacking strip, (2) tack strip, and (3) flexible tack strip (see Glossary).

Installing rigid tacking strip

Rigid tacking strip is used anywhere an outside panel is to have a straight top edge. Units with straight arms, straight top backs, bands with straight edges, and almost all skirts will use the rigid tacking strip. Follow this procedure:

1. Locate and spot-tack the OA panel at the front end of the panel (FIG. 16-1) and work toward the back of the unit, lightly stretching to take out any wrinkles. Notice the extra material at the rear. This will be tacked to the back upright. The outside back (OB) will then be installed over the top of it.

   A second approach is to tack the center of the OA in place, then work toward each end (FIG. 16-2), again lightly stretching to remove any wrinkles. The extra material at the front end is tucked under and blind-tacked using a tack strip, as explained below.

   Figure 16-3 shows an OA stay tacked in place. Notice that the staples are 3–5-inches apart. This is all the stapling that is necessary at this point. *Important:* Keep the staples a little way from the very top so that when the tacking strip is applied, all stay staples will be concealed.

2. Obtain an appropriate length of tacking strip. Many upholsterers use a ½-inch lightweight tacking strip that can be torn to length with the fingers. This is often torn a couple of inches long, then cut or torn to final length when installed to the unit. The heavy-duty tacking strip (FIG. 16-4) is best cut with dikes. Shears may be used but are not quite as comfortable to use,

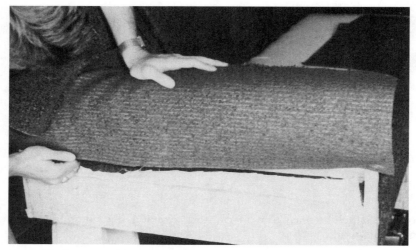

16-1 Aligning an outside arm panel, starting at front, then stretching toward back.

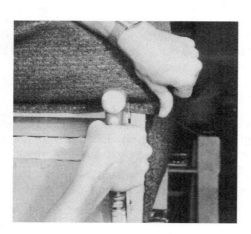

16-2 Starting OA panel at center and stretching toward each end to complete stay-tacking.

and the heavy-duty tacking strip dulls shears prematurely. (Don't try to tear this.) The *final* length of either light weight or heavy-duty tacking strip should be about ¾-inch short at each end where other blind-tacking material is to be used going down one or both sides.

3. Lay the tacking strip along the stapled edge and working from the center (FIG. 16-5), staple in place. Cut ends to allow for perpendicular tack strips as necessary.

Stapling (or tacking) close to the top edge of the tacking strip (FIG. 16-5) will give the best results—keeping that top edge tight against the frame.

If heavy, thick fabric is used, or there are multiple layers (as with a welt plus the inside panel plus the outside panel), it is advisable to install the staples at a diagonal, as shown in FIGS. 16-6 and 16-26. Notice the angle of

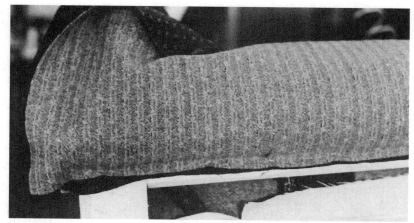

16-3 An outside arm panel stay-tacked in place.

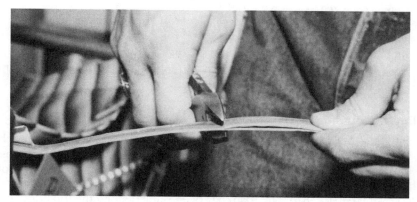

16-4 Cutting heavy tacking strip with dikes.

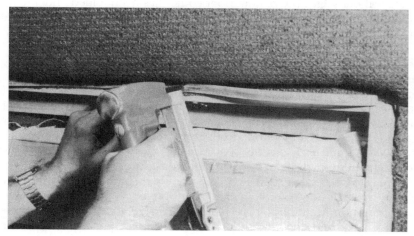

16-5 Begin stapling tacking strip at center, working toward ends.

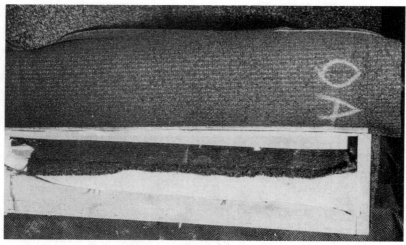

16-6 Tacking strip for OA in place. Notice angular placement of staples, panel identification, and orientation mark (chalk mark at extreme top right of photo.)

the gun in FIG. 16-26. This will prevent the inside edge of the tacking strip from "pooching" out and creating an objectionable ridge in the panel when it is pulled down.

Padding the outside arm

Frequently it is desirable to pad the outside panels (arm and back). Padding the outsides gives a more plush, softer appearance and feel to the whole unit. On units having a low profile, like the one shown in FIGS. 16-7 and 16-8, attaching a layer of ¼- or ½-inch soft foam will do nicely. No additional base is needed because of the narrow span to cover (see FIG. 16-6). The foam has been stapled along the top edge (FIG. 16-7) over the top of the tacking strip. Notice that it is stapled sparingly, just enough to hold it in place without sags. (The bottom front corner has been folded under temporarily just to identify clearly what is happening.) Note also that the foam extends about ¼ inch beyond the bottom. This will give the edge a soft appearance and feel. With soft foam, it compresses to the point that no objectionable ripples will be created.

This is one of the few cases where a padding material will go around an edge onto the surface to which it will be stapled. A few staples along the front and back uprights and bottom rail will hold the foam in place. Spray adhesive can also be used.

To finish the outside arm panel, pull the panel down near the center and staple (FIG. 16-8). Notice the abundant material for pulling. This is much better than not having enough. Stretch the bottom, front corner on a diagonal, as indicated by the arrow (FIG. 16-9) and tack on the bottom. Pull down the fabric between staples, aligning the pattern, and staple in place along the bottom of the rail.

If there is to be a rolled edge along the front of the arm post or stump (as in this case), staple a small ridge of cotton along the front edge and form the roll to make a smooth transition from the roll created by the inside-arm installation. Figure 16-10 shows the small rolled edge being formed with the OA panel. A narrow finishing *panel* will be installed later to complete the front of the arm stump.

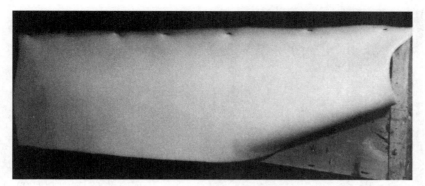

16-7 Stapling a ¼-inch foam pad in place for OA.

16-8 Stretch and staple bottom of OA, working from center.

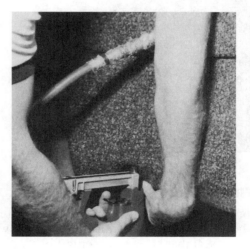

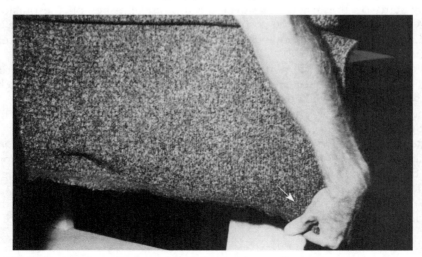

16-9 Stretch diagonally (arrow) to fit front and rear of panel.

16-10 Making rolled edge with OA panel along lower portion of front arm stump.

COVERING A ROLL-OVER SIDE (SEAT)

Although this cannot really be classified as an arm, it comes closer to that than any other part. A roll-over edge is a very low, padded, rather solid end to a bench-type seat that has a low, vertical back. In this case, the outside "arm" panel actually starts on the inside, rolls over the top, and finishes off the outside as well. A useful trick of the trade is shown in FIG. 16-11—a Y cut made for the back post. Can you see the very low elevation from the decking that is being created by the roll-over edge?

In this case, the tab will remain upright and be stapled to the post rather than tucked down. This will create a fabric "bridge" that will make it impossible to have frame wood showing when the back panel is pulled down over it.

The rear flap is also left out and will be stapled in place to create another fabric bridge (FIGS. 16-11 and 16-12).

After the bottom center is stapled in place, the corners are stretched diagonally and stapled. In FIG. 16-12, the fabric is being stapled to the *side* of the rear upright rather than the back because a narrow panel will be installed to finish off the upright. The OB will also be stapled near the same point, as shown in FIG. 16-12. Notice the upward curve to the fabric pattern between the back post and the center. This will be removed when the portion between the staples is pulled into place and stapled.

16-11 Y cut to fit OA to back post.

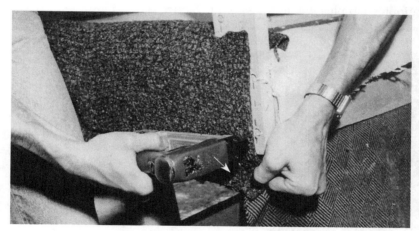

16-12 Diagonally stretching (arrow) to fit bottom rear of OA.

INSTALLING FLEXIBLE TACKING STRIP

Flexible tacking strip is used to attach panels with curved edges. In the earlier days of the industry, all of this work was sewn on with the blind stitch or applied with decorative tacks. All three methods are still in use, and each has its specialty.

Blind stitching is about the only method that can be used where there is no frame immediately beneath the locations where the attachment is to be made. It is extremely economical (in materials), requiring only sewing thread. Stitching requires a bit more craftsmanship and time than the other two methods.

Decorative tacks are used to add a distinctive decoration to the installation and where the presence of the raised heads will not be objectionable. One such objection would be using decorative tacks to apply the backs to dinette chairs that will be used near walls or against wood or painted furnishings. The tacks will mar such surfaces significantly. A requirement for using tacks is that there must be framing material in the area of the attachment. Of the three methods, this is the most expensive in materials.

Flexible tacking strip (brand names Curve-Ease and Pli-Grip) finds its specialty application anywhere curved surfaces are involved and visible tacks are undesirable. Applying flexible tacking strip can be accomplished with relative speed and ease. Material cost is almost negligible. As with tacks, framing beneath points of attachment is a must. Figure 16-13 shows a view of the back of the tacking strip with the pronged tab folded over as it would be in an actual application. Two prongs are visible on the left side; these will grip the fabric as they are pressed toward the frame of the chair. The rear tab (the one to the front in the photo) has a hole punched in it through which a tack or one leg of a staple will pass, attaching it to the frame. The cloth measuring tape has been inserted between the two tabs to show their dimensions. The pronged (gripping) tab is ⁷⁄₁₆ inch, the attaching tab is ⅜ inch. Leave a space a little more than ¹⁄₁₆ inch between the edge of the attaching tab and where the finished edge of the fabric is to be located. This is to allow for that additional ¹⁄₁₆ inch of the gripping tab plus the thickness of the fabric.

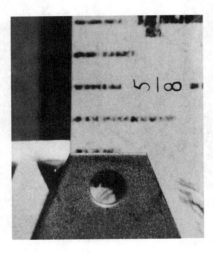

16-13 View of flexible tacking strip showing "grippers" (left), hole for attachment, folded dimensions, and overlap.

The following procedure is used when installing flexible tacking strip:

1. Locate the flexible tacking strip about ¹⁄₁₆ inch from the edge to be finished. Bare wood was used in FIG. 16-14 to illustrate this concept.

2. Continue attaching the strip to within 2 inches of the point it is to terminate and cut it to the needed length with a pair of tin snips (FIG. 16-15). Aviation snips, as shown in FIG. 16-15, are preferred by most upholsterers. Figure 16-16 shows the flexible tacking strip applied to the outside arm of a recliner.

16-14 Attaching flexible tacking strip. (Note particularly the position of the attaching tab and the staple gun.)

16-15 Cutting flexible tacking strip to length with tin (aviation) snips.

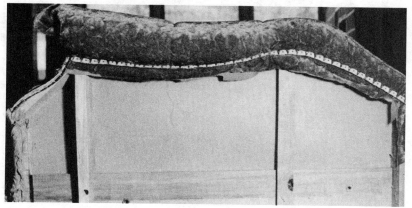

16-16 Flexible tacking strip in place, ready to contour OA panel.

3. Align, mark, and trim the panel.
   ~Push all the pronged tabs of the tacking strip about half closed so they are at about a 45-degree angle.
   ~Lay the OA (or OB) panel over the area and pin or stay-tack in place so it will not move during the marking. Figure 16-17 shows a panel that has been stay-tacked to the bottom of the recliner, stretched and stay-tacked at both front and back (not shown), tucked behind the bent ears of tacking strip (distinguishable by the dark line just above the chalk mark), pinned at the top, then marked with chalk. The best way to mark the contour is to rub the side of the chalk along the edge of the metal.

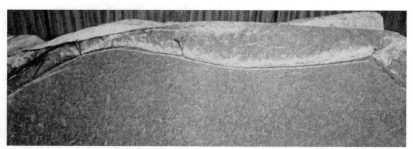

16-17 OA panel pinned (top) and stay-tacked (ends and bottom), tucked behind flexible tacking strip and marked for trimming to contour.

   ~Remove the skewers and stay tacks and trim the fabric about $\frac{7}{16}$ inch beyond the mark, as shown in FIG. 16-18. A normal ½-inch allowance gives too much material to tuck beneath the flexible tack strip.
4. Relocate the fabric over the flexible tack strip, making sure the marked line exactly matches the contour of the tacking-strip edges.

16-18 Trimming OA panel to contour, ⅜ inch beyond chalk mark.

5. Hook the fabric onto the toothed segments, (FIG. 16-19). Most of this work will be done by pressing the fabric around with the fingers.

A *skewer* is a handy helper to move the fabric around accurately. The operator in FIG. 16-19 is using one.

Keep the panel taut during the tucking-under process to eliminate those unwanted "puckers" that will otherwise appear later.

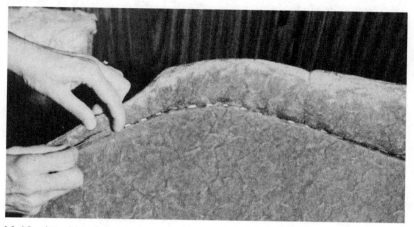

16-19 Attaching OA panel to grippers using skewer to help tuck the edge in.

6. Press the gripping tabs over against the frame. Fabric slippage can be prevented by pressing toward the tacking strip just inside the metal edge with one hand, as done by the left hand of the operator in FIG. 16-20, while pressing the nearby tabs over with the other thumb. The panel to the right of the operator's right arm has been folded into place.

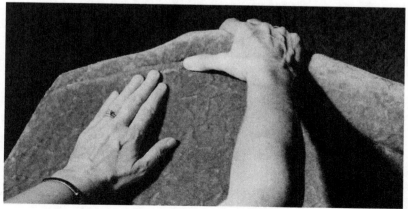

16-20 Maintaining tension on panel while folding tabs down.

7. Seat the tabs firmly against the frame. One handy way to do this is to use a piece (about 2½ × 4 × ¾ inches) of close-grained hardwood, like maple, cherry, or birch to tap the tabs firmly in place. A very handy short dowel "handle" was added to the one in FIG. 16-21.

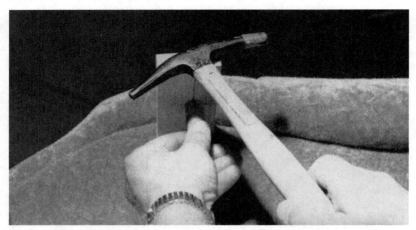

16-21 "Setting" flexible tacking strip tight against chair, using the end grain of a hardwood block.

8. Stretch the bottom and ends of the panel as necessary and staple in place. The finished job should look something like FIG. 16-22.

16-22 Completed outside arm using flexible tacking strip.

THE OUTSIDE BACK

Outside backs are installed in much the same manner as the outside arm:

1. Start with the top, *center* the pattern (FIG. 16-23), and spot-tack in place. Figure 16-24 shows the center staple being set.

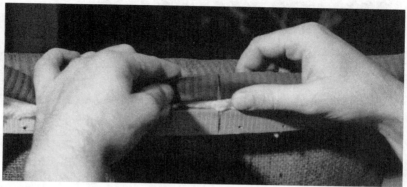

16-23 Aligning marked centers of OB panel and top rail.

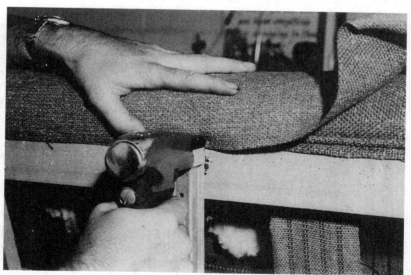

16-24 Stay-tacking OB panel in preparation for tacking strip.

2. Stretch the fabric sideways and spot-tack along the top to the sides of the unit. Notice that on a straight-topped back rather long spans (FIG. 16-25) can be taken when making the stretch. A few staples will be set later between those shown just to hold the edge in proper alignment.

16-25 Example of length of run in place-tacking OB to couch. Note centerline notation on frame, right of photo.

3. Measure, cut, and install the tacking strip as indicated earlier in FIGS. 16-4 through 16-6. Figure 16-26 shows the diagonal stapling using the heavier tacking strip. (It can be identified as heavy tacking strip because of the greater than normal spacing between the staples. When using lighter-weight tacking strip, staples will be rather close together.)

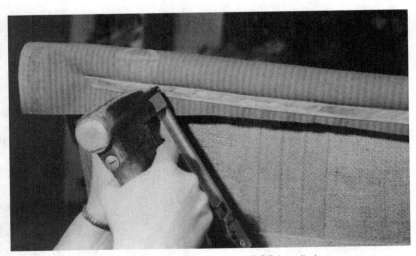

16-26 Diagonally stapling tacking strip for top of OB installation.

4. Measure and cut two pieces of *tack strip* to length, as indicated in FIG. 16-27.
   The top of the strip should be in line with the top edge of the tacking strip, and the bottom will end just short of the bottom of the seat rail. (The left of FIG. 16-27 is actually the bottom of the couch.)
   Relocate and cut the tack strip if necessary so there will be a tack near both ends. This helps eliminate any tendency for the ends to "lift."

16-27 Measuring and cutting tack strip to length.

5. Trim the OB panel to size. Best results will be achieved if the OB panel is trimmed so there is about ¾ inch to be rolled under. (The rolling under will be explained shortly.)

6. Locate the tack strip in place. The inside edge of the strip (with the tacks pointing outward) is to be along the inside edge of the welt (if welt is used, as in this case), as illustrated in FIG. 16-28.

16-28 Locating tack strip alongside of OB.

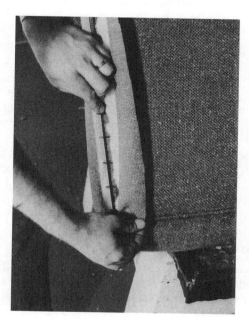

7. Set the tack strip. Hold the strip carefully in position while lightly pulling the fabric over the tacks and setting it partially onto the tacks, (FIG. 16-29).

   Push the fabric over *all* tacks partially (FIG. 16-29), then go back and set the fabric all the way to the strip base. This will avoid creating puckers that occur when trying to set the fabric all the way for each tack.

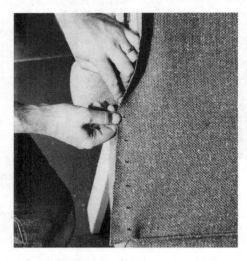

16-29 Lightly stretching OB panel and pressing it onto tack strip.

8. Roll the tack strip and fabric inward. It should require a slight stretching to get the, now, *outer* edge of the strip to coincide with the *inner* edge of the welt. If it does not take a little effort to tighten the fabric or it cannot be stretched far enough, remove the strip and start over, making adjustments as necessary. The fabric must be *taut* but not so taut that fingernails are broken trying to pull it into place.

9. Set the tack strip. Ideally, the tacks will be angled so the points are slightly outward from the heads, as in FIG. 16-30. When the strip is set, the strip will tend to tighten the fabric just a little more. Use the side of the tack hammer (FIG. 16-31), a hard white rubber mallet, or a rawhide mallet to set the strip. *Caution:* The major force must be centered over the heads of the tacks (FIG. 16-31), not between them; otherwise the strip will be destroyed and the fabric damaged.

   Occasionally, when using the tack hammer, the hard steel side may cut some fabrics because of the pressure needed to drive the tacks into harder woods like oak. To avoid that, place a couple of layers of scrap fabric over the strip to finish setting the tacks. Figure 16-32 shows one corner of a properly finished outside back.

Offset backs

Occasionally a unit will have a little jog or offset along the sides of the back, and a single straight tack strip cannot be used. Use two pieces for each side. A little problem solving is needed for each case, but most cases are similar.

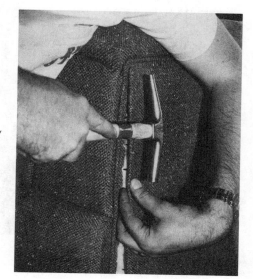

16-30 Tack strip rolled under, ready to set tacks.

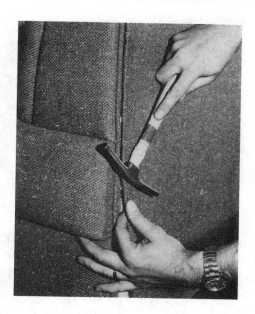

16-31 Apply driving pressure directly over tack heads (not between them).

1. Cut and install the top tack strips. These should end where the jog takes place. It will probably be necessary to make a short diagonal cut into the fabric at this point to permit the top flap to be folded under with the strip and another flap to fold under to create a finished edge at the jog. Figure 16-33 shows a chair back with the top tack strips installed.

2. Cut and install the bottom two tack strips. Figure 16-34 shows the finished back with a jog.

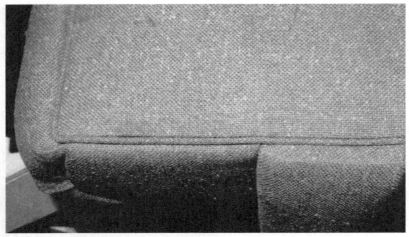

16-32 Finished end of OB panel on sofa.

16-33 First steps in installing an OB panel with a bottom width wider than the top (offset).

16-34 Done properly, a change in width of the OB panel is nearly invisible.

17
Cushion making

This chapter will deal with cushions of a reversible or loose styling. No attempt is made to cover the particulars of cushions attached directly to the unit.

All styles of cushions are started in the same manner—by determining the appropriate size and shape. To do this, one of three basic methods will generally be used, whether the cushion is a box, knife-edge, or waterfall style: (1) *patterns*, usually most appropriate for producing multiple units of the same style and size; (2) *measuring*, to obtain length and width for cushions having straight regular sides; (3) *tailoring*, the most accurate method to match cushions to irregular or curving perimeters.

Patterns are made and measuring and tailoring done after the deck and inside arms and back have been completely padded and covered. That is the only way the proper dimensions can be determined reliably.

Once the orientation of the nap and cover pattern have been determined, a few preliminaries must be attended to before getting to cushion construction. These particulars are presented in TABLE 17-1. The considerations are listed in the left column, the specific treatment under each cushion style. The knife-edge cushion is made in two basic styles: (1) the midseam on the front and sides (for seat cushions) or top and sides (for back cushions); (2) the midseam only on the front (for seat cushions) or top (for back cushions). These two styles are specified in TABLE 17-1.

TAILORING

For all three cushion styles (box, knife-edge, waterfall), one oversized thickness of the main panel is smoothed onto the deck area, face side up, as shown in FIG. 17-1, with the front oriented properly. Smooth out the panel so that the back and sides are tucked under about even distances and let the excess extend over the front crown. Now the tailoring process can begin.

1. Mark the perimeter with chalk (FIG. 17-1). This is done by holding the chalk vertical with the back or arm and marking on the *face* side of the cushion, as shown. The side of the chalk should be in line with but not pushing into the padded back and inside arm at a point approximately half the cushion thickness from the deck level.

2. Mark the cushion line, where more than one cushion is to be tailored for the unit (refer to FIG. 17-1, far right side). These cushion lines should have been marked on the decking and crown of the seat panel, prior to tailoring cushions.

Preparation & consideration	Box cushion	Waterfall cushion	Knife-edge cushion
Number of panels	2	1	2
Pattern or nap orientation	Forward on both.	Forward on top, will be reversed on bottom.	Forward on both.
Rough cut size of cushion panels	Max. width +2" Max. length +2"	Max. width +2" Max. length +2" + boxing width.	Max. width +2" Max. length + boxing width.
Boxing style	Seams parallel to top and bottom (all 4 sides).	Seams parallel to top and bottom at sides & back. (Rounded nose at front of sides.)	TWO STYLES: Notched corners: No boxing, seams & welt at middle of sides. Tucked corners: Seams parallel to top and bottom at sides & back. (Rounded nose at front of sides.)
Seam & side allowances	Back, sides & front: ½"	Back and sides: ½" Front: width of finished boxing.	TWO STYLES: Notched corners: Front, back & sides: ½" + ½ cushion thickness. Tucked corners: Front: ½" + ½ cushion thickness. Back & sides: ½"
Special tucks	None	TWO STYLES: Square corners: None Rounded corners: ¼" at front seam lines.	TWO STYLES: Notched corners: None Tucked corners: ¼" at front seam lines.

3. Mark the seat crown at both sides (FIGS. 17-2 and 17-3). If a waterfall
cushion, lay off two additional marks from the crown mark that measure
half the cushion thickness each. Figure 17-4 shows these markings for the
standard 3½-inch cushion. Fold the panel on the centerline (in this case, the
8¼-inch mark, FIG. 17-4), aligning the weave or pattern and staple together
(FIGS. 17-5 and 17-6).

Caution: Be sure *not* to fold the panel at the seat-crown mark. Staple
inside the seam lines.

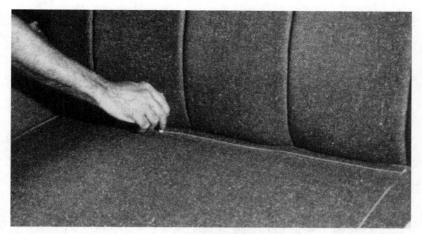

17-1 Tailoring cushion panel: marking centerline (right side) and perimeter for one of two waterfall couch cushions.

17-2 Marking seat crown at end for waterfall cushion, part of tailoring operation.

4. Trim ⅜ inch for seam allowances rather than the standard ½ inch (FIG. 17-7). The reason for this ⅛-inch deviation is to allow for the blousing-out of the cushion sides when *over*stuffed with foam. If you allow for the full ½-inch seam, the cushions will be too tight in the unit and may even buckle. Figure 17-8 shows a cushion panel trimmed.

5. For waterfall cushions, make tucks at crown marks (6½- and 10-inch markings, FIG. 17-4) so that the fold faces downward the centerline mark, as shown in FIGS. 17-9 and 17-10. The tucks should be approximately ¼ inch and terminate at the crown marks.

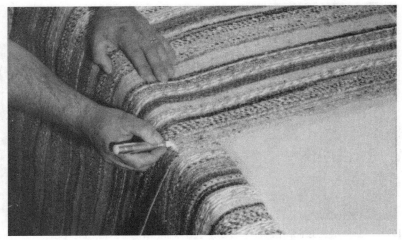

17-3 Marking seat crown at cushion line for waterfall cushion.

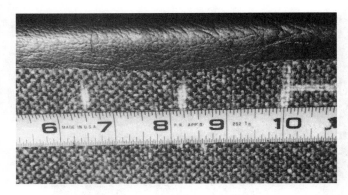

17-4 Markings for "standard" 3½-inch thick waterfall cushion: 8¼-inch mark is the centerline of the waterfall cushion, the 6½-inch and 10-inch marks are the crown marks (or cushion thickness).

17-5 Stapling waterfall cushion panel (folded along the centerline or 8¼-inch mark shown in FIG. 17-4) for trimming.

17-6 Place staples inside the seam line to clear blade when cutting.

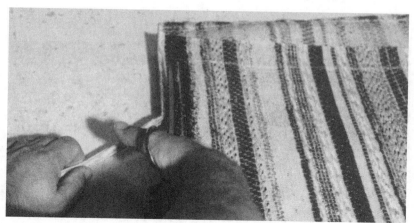

17-7 Trim excess fabric ⅜ inch from tailored seam lines. *Do not trim the standard ½ inch; that would make cushions too wide.*

17-8 Cushion panel trimmed within ⅜ inch. A standard ½-inch seam allowance will be taken when sewing cushion.

17-9 Fold tuck to each crown mark and staple (inside sewing line), preparing the panel for sewing.

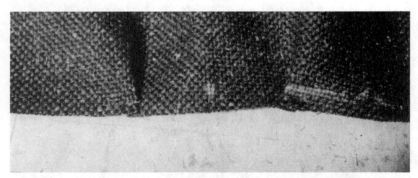

17-10 Tucks stapled, ready to be sewn. (Note that the tucks face toward the centerline.)

6. Notch the centerlines of the cushion and boxing panels, as indicated in FIGS. 17-11 and 17-12. Round the front corners of both boxings as indicated in FIG. 17-12. Tailoring for the cushion is now complete.

CONSTRUCTION TECHNIQUES

Most cushions built for the furniture industry will use a zipper to facilitate stuffing and speed up closing the cover. In earlier times, many cushions were closed on the rear seam by hand-sewing, using the blind stitch. Convenience and speed are the major reasons for including zippers.

17-11 Notch centerline of cushion panel to align with boxing center.

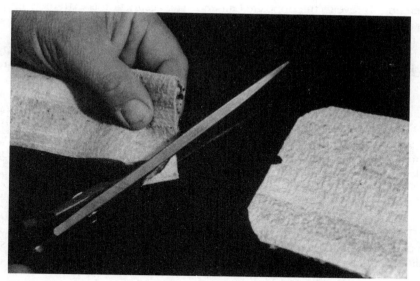

17-12 Rounding corners of boxing panel; centerline has been notched.

Figure 17-13 shows a strip of number 2 brass zipper stock and five glides—preferred for most furniture upholstering. A dime has been included for size comparison along with a glide to a larger zipper. (The larger size would commonly be found on sleeping bags, tents, heavy jackets, etc.) Figure 17-14 shows a close-up view of the two opposite ends of a piece of zipper stock. The fabric has been trimmed as closely as possible to permit viewing the two in close proximity. Point A shows the

17-13 Zipper stock and glides (dime included for size comparison).

17-14 Close-up of opposite ends of zipper stock: (A) cupped end; (B) ball end.

cupped end; B reveals the opposite ball-end. This distinction is extremely important to installing the glide properly. The real problem is that the glide can be put on either end but will zip closed only when facing the proper direction.

The ball end of the zipper stock must point in the same direction as the tapered end of the glide, as shown ing FIG. 17-15. To install the glide, slip the square end down over the zipper stock (FIG. 17-16). Applying a light downward pressure on the glide with one finger (A), grasp the extreme top edges of the fabric and by rolling the hands away from each other at the top (B–B), attempt to separate the metal segments. This sometimes takes a little wiggling. Raise the "pull" away from the glide body so the locking nib on the underside does not engage with the stock and prevent movement. Notice in FIG. 17-16 that the pull is resting on the right thumb. That is to keep the locking nib from engaging.

Once the glide has been installed onto the stock, slide it up and down to make sure it works smoothly. If it will open the zipper with ease but will not close with the same ease, chances that are the glide has been installed upside down (from the wrong end). Take it off, examine the ends carefully to ensure that the glide point is facing the same direction as the ball end of the stock, and try again. If it will not move, just slide the glide all the way off the way it will move, opening the zipper completely.

17-15 Pointed end of zipper glide must be oriented the same as the ball end of the zipper stock, or glide won't work.

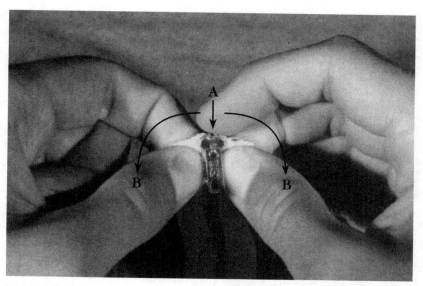

17-16 Installing zipper glide onto stock. Arrows show direction of simultaneous pressures.

Now the zipper is in two pieces (FIG. 17-17). That can be remedied with ease. Start at the cupped end and roll the interlocking segments together, as indicated in FIG. 17-18. By rolling from the bottom up, the segments can be felt making their interlocks. To speed up the process, once the bottom few segments are locked, place the zipper on a flat, smooth surface and with a pinching-rolling action interlock the segments, as illustrated in FIG. 17-19.

17-17 Zipper stock that has been separated.

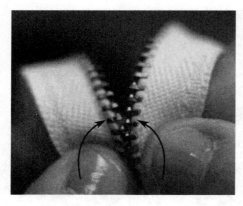

17-18 Freehand method of interlocking separated zipper stock.

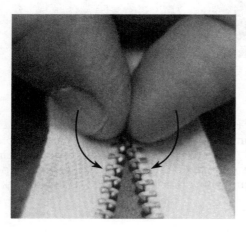

17-19 Smooth surface speeds up zipper-stock interlocking. (Arrows indicate a rolling, pinching motion of the fingers.)

A faster way to close completely the longest of zippers in a matter of seconds is to use a zipper glide that has been opened (it is no longer any good as a glide but makes a super closing tool) and slide it along the top of the stock that is resting on a smooth, hard surface, as shown in FIG. 17-20. Not all zipper materials will work this way, but most upholstery stock does.

SEWING SEQUENCE (WATERFALL CUSHION)

The following sequence is suggested for sewing most styles of waterfall cushions. The one pictured in the sequence is a contemporary style, having no welt. However, as an aid for those who want to include a welt, the point at which the welt would be sewn on is included.

1. Sew the first half of the zipper panel to the zipper stock (FIG. 17-21). The ½-inch seam allowance has been folded under and the folded edge aligned with the center of the zipper stock. Proceed with the second half of the zipper panel so that the two folded edges meet over the center of the metal interlocks of the zipper (FIG. 17-22).

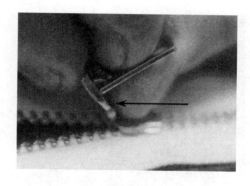

17-20 Superfast method to interlock separated zipper stock. (Straight-line run with opened zipper glide does the job.)

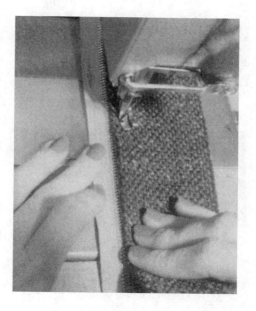

17-21 Sewing folded first half of zipper panel to zipper stock to create a concealed zipper.

2. Sew the boxing (square end) onto the zipper stock (FIG. 17-23). The excess zipper stock is easy to see, and if you look closely, the darker edge of the face side of the zipper panel is barely visible along the right edge of the boxing.

    Caution: Be sure that the zipper glide has been installed before completing this step (see FIG. 17-16). Be sure that the fabric panels are face-side to face-side.

    The excess zipper stock visible to the right of FIG. 17-23 will be trimmed off after the boxing has been attached. Backstitching over the zipper stock (which is easy to see in FIG. 17-23) will reduce the tendency for the zipper to break through the thread.

17-22 Sewing second half of zipper panel to zipper stock.

17-23 Sewing boxing to zipper panel (face sides together).

3. Locate and mark the center of the zipper panel (FIG. 17-24), both sides. These marks are to be aligned with the center marks made in the cushion panel. Figure 17-25 shows a striped boxing sewn to the zipper panel. Notice that the stripes of the zipper panel will be aligned with the pattern on the cushion panels and that the boxing panel will have the pattern facing forward (the "down" direction). Figure 17-26 shows a seamstress marking the excess stock from the zipper panel for trimming. (Some operators will make the zipper panel extra wide and trim it to the boxing width after sewing the boxings to the zipper.)

17-24 Notching centers of zipper panel to align with rear center of cushion panel.

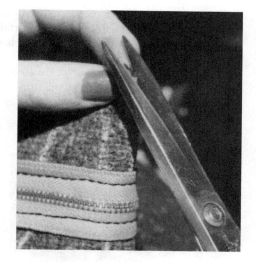

17-25 Trimmed boxing and zipper panel.

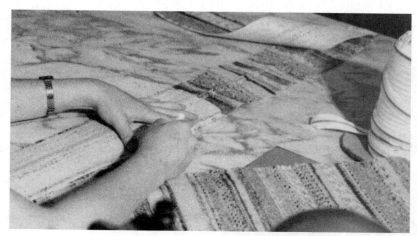

17-26 Marking oversized width of zipper panel for trimming prior to sewing to cushion panel.

4. If *welt* is used: Sew welt to cushion panel so that the joining seam will be at the rear of the cushion. (On waterfall cushions, locate the single joining seam at the rear bottom.) To join welt so the seam has the least bulk, follow this procedure:
   a. Leave about 4 inches of the welt free at both ends.
   b. Unpick the ends of the welt strip (FIG. 17-27) so the fabric can be laid out flat where they will overlap, at least 3 inches from either end. This will be illustrated later.

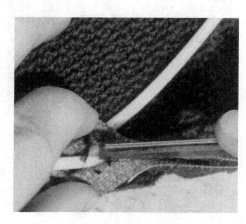

17-27 Open both ends of welt at least 3 inches beyond overlap point.

   c. Stretch the cushion panel and the overlapping welt panels taut to locate the center point (FIG. 17-28). Mark both welt strips at this point as shown.
   d. Locate opened welt strips face-to-face at center of chalk marks (FIG. 17-29). Push a skewer through the centers at the overlap point.
   e. Place the skewered assembly on a flat surface and rotate the strips to be 90 degrees to each other, as illustrated in FIG. 17-30.

17-28 Mark welt at point of overlap.

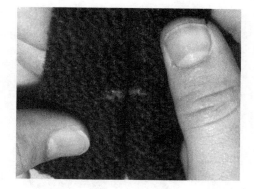

17-29 Align centers of overlap marks (fabric is face-to-face).

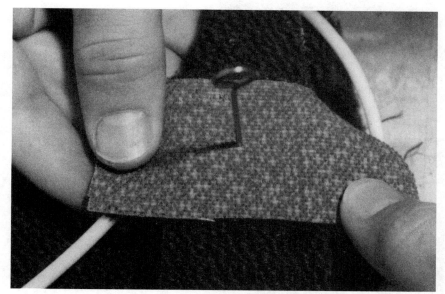

17-30 Pinned centers (top strip has been rotated 90 degrees to bottom).

f. Hold the assemby in this position, remove the skewer, and sew across diagonal corners so that when opened the strip will lie in line and not overlapping itself (FIG. 17-31). To check this orientation, hold fingers along the line of one of the diagonals and lay over the top strip. It will lie on top of itself or in line; if on top of itself, sew the other diagonal.

g. Trim the excess close to the seam (FIG. 17-32) within ⅛ to ¼ inch to reduce unnecessary bulk. Open what is left of the flaps, shown almost beneath the fingers in FIG. 17-33, to minimize excess bulk.

h. Cut cording to make a butt joint at a spot not directly beneath the joint in the stirp (FIG. 17-33).

17-31 Sewing across diagonal.

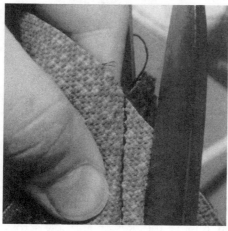

17-32 Trim off excess ⅛ inch to ¼ inch from seam.

    i. Finish sewing the joined welt to the cushion panel (FIG. 17-34; the arrow points to the diagonal seam where the two ends of the welt are joined—in this fabric, tough to see).

5. Attach the center front of the boxing to the center mark of the waterfall cushion panel, as shown in FIG. 17-35. Sew the boxing to just around the rear corner of the cushion panel, then align the zipper-panel center notch with the rear-cushion-panel center notch (FIG. 17-36). Leave the machine needle in the *down* position while making this adjustment to ensure no movement from the previous stitch.

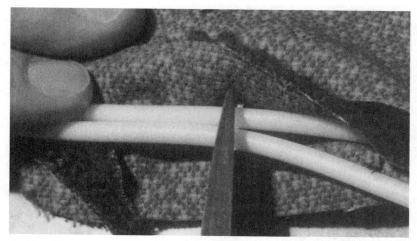

17-33 Opened seam has been flattened. Cut cording at overlap point that is not directly over seam.

17-34 Complete sewing of joined welt to cushion panel. Arrow points to the joint of the seam.

6. Now go back to the corner. Fold over the excess boxing so the zipper panel will lie out straight, as shown in FIG. 17-37. *Caution:* Keep the center notches aligned and sew straight over the top of the folded material. This creates a *zipper pocket*. Sew to just beyond the center notches and terminate sewing along this direction; cut the thread.

7. Go back to the front of the boxing and prepare to sew up the opposite side as indicated above. Figure 17-38 shows the seamstress aligning the panels for this final sewing step. When reaching a previously sewn area, always backstitch to prevent the stitches from opening later.

Your cushion is sewn. Now stuff it with foam.

17-35 Align boxing center with front of cushion center (for a box cushion) and sew around one side, ending about 3 inches short of the rear corner.

17-36 Align zipper-panel center with rear-cushion-panel center.

17-37 Folding over boxing to create a zipper pocket.

17-38 Preparing to sew second half of boxing to cushion panel.

18
Skirts & cambric

SKIRTS

Skirts come in four basic designs: *flat panel*, probably the most popular, (FIG. 18-1, left), *gathered*, *box-fold* (FIG. 18-1, center), and *accordion-fold*.

Skirts are used to conceal platforms and legs from view and to give the unit the appearance of resting on the carpet. There is a third reason for using skirts: The customer likes the looks of them. Figure 18-1 shows three types of furniture with skirts: a couch, plaid pattern; a swivel rocker (center); and a recliner rocker (right).

The recliner rocker (FIG. 18-2) uses the flat panel skirt to conceal the platform base and to disguise the fact that it is a rocker. This style of skirt is by far the most popular. It takes less cover fabric to make, gives a smooth elegance to the unit, and is the easiest to figure out. Usually, when a flat panel skirt is selected, a *kick pleat* will be used to cover around corners and sometimes where individual panels meet at the front or backs of couches. A kick pleat is merely a small panel of fabric used to cover openings where the regular panels come together.

18-1 Three styles of skirts.

18-2 Skirt on bottom of recliner foot rest matches sides of chair.

18-3 Skirts pinned up to reveal three construction techniques.

The three skirts have been pinned up in FIG. 18-3 to show the totally different construction of each. The close-up view of the recliner rocker (FIG. 18-4) shows the most formal and highest-quality skirt. It has a liner sewed to it so that the cover fabric is rolled around to be a part of the backing. This style gives the skirt a "full" feel and appearance. It is also the most time-consuming to make.

18-4 A skirt design that has liner sewed in all around.

18-5 A closer view of the swivel rocker reveals a distinct mistake. Never have a seam visible from the front.

A closer look at the swivel rocker (FIG. 18-5, right) exposes something that should never happen. Can you see the seam in the front skirt? Any seams that *must* be made in skirts should be located at the back of the unit and be hidden within one of the folds, not exposed in the center of one.

Figure 18-6 shows the underside of this skirt and the simplified method of construction. The bottom edge of this style of skirt is finished off by merely spraying the back side with fabric and foam adhesive, folding up the edge about ¾ inch, and gluing the bottom to itself. This gives a finished edge to the bottom. One disadvantage of this style is that the bottom edge is now twice as thick as the rest of the skirt, which will prevent the bottom edge from folding as neatly as the rest of the skirt.

18-6 Simplest method of skirt construction: bottom edge is folded over and glued (foam and fabric adhesive) to itself.

The third style of skirt is shown in FIG. 18-7. For this style, a flat panel, a piece of liner material normally a bit stiffer than the cover fabric, is glued to the back of the skirt. This adds stiffness to the skirt and keeps the individual panels relatively straight and smooth for a long time.

Figure 18-9 shows how the kick pleat is stapled onto the unit at a corner. Notice that the kick pleat is stapled on top of the flat panels. When the skirt is in its normal downward position, the kick pleat will be beneath the flat panels. Notice the white backing material. This is either sewn to the cover fabric or glued on.

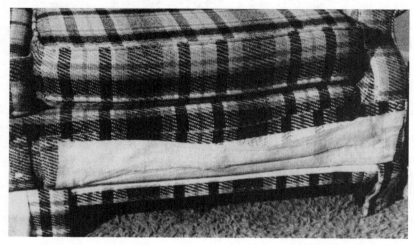

18-7 A third method of skirt construction: a stiffener-liner is glued rather than sewn to back.

Figure 18-7 shows a kick pleat used to cover the joints between two flat panels. In this case also, the kick pleat is stapled on top of the upturned skirt so that when they are all in their normal downward position, the kick pleat is "beneath" the skirt panels.

A closer look at the back side of this style skirt (FIG. 18-8) permits you to see that the bottom edge is glued like the swivel rocker and the stiffener material glued to the back afterward. Kick and corner pleats have the same construction (FIG. 18-9). A kick pleat spanning the union between two flat panels will be attached, as on the front of a couch, in the same manner as the corner pleat featured in FIG. 18-9.

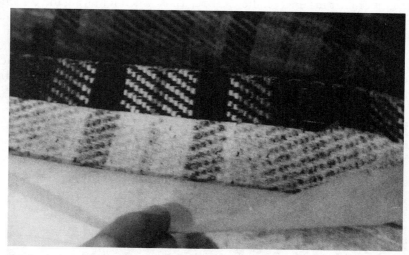

18-8 A closer view of skirt featured in FIG. 17-7, showing how skirt is folded and glued, then liner glued on top. After a time (17 years in this case), liner comes unglued rather easily.

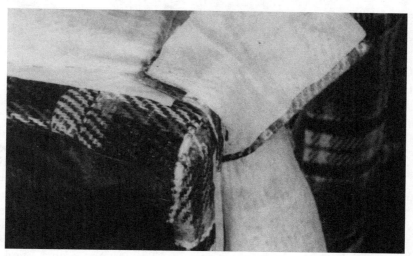

18-9 A corner pleat of the same construction as the flat panels on the plaid patterned couch.

CAMBRIC

Cambric is a lightweight, close-weaved fabric used to finish off the bottoms of chairs and couches. In the earlier days of the industry when stuffing and padding materials were made from animal hair, sisal, tow, and coarse cottons, the constant movement of the padding would break off little pieces or loosen fibers so they would fall on the floor. To prevent the floor from constantly having a clutter of debris beneath the furniture, cambric was used to catch the material.

In reupholstering older furniture pieces today, it is not uncommon to find a significant deposit of filler material on the cambric. Thus, when removing the cambric from these older pieces, the first thing many upholsterers will do is vacuum up all dust and stuffing particles. Most of the cambric used in modern upholstering is used primarily to finish off rather than act as a dust catcher. And most cambric used on the bottoms of chairs and couches has a *face side* and a *back side*. The face side is shiny; the back side, rather dull (FIG. 18-10). The face side goes to the outside, as with any other fabric. Although rather simple in installation, a few tricks may prove helpful.

18-10 View of cambric showing outside (shiny) and underside (dull) of the fabric. (Not all cambric is black.)

1. Measure the major dimensions of the bottom (FIG. 18-11) to determine the size to cut. Allow ½ to 1 inch per side to be folded under. More or less can be used when it is more convenient or necessary since all that is really needed is enough to make a finished edge. The allowance suggested is a quick way to ensure sufficient coverage without taking a lot of time for calculation.

2. Anchor front and back centers, tucking a flap *under* to create a finished edge (FIG. 18-12). (This is just the opposite of *burlap* installation, where the flap is turned toward the outside to prevent tear-out. With cambric, a finished edge is the concern, not tear-out.) For best appearance, staple near the edge of the fold, within ⅛ inch.

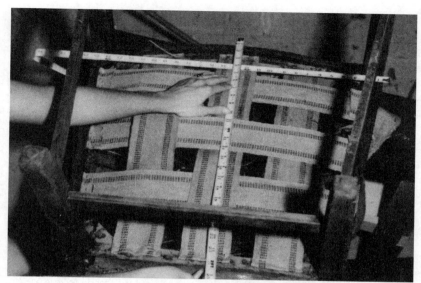

18-11 Measuring major dimensions to get size for cambric panel.

18-12 Locate, fold under edge, and staple at all centers first. Then work toward and finish up at corners. Double-folding corners is faster than trimming out excess fabric.

3. Cut (diagonal or **Y**, as required) and fit neatly around legs that are a permanent part of the chair (as in FIG. 18-13) and staple to the bottoms of the rails whenever possible. (See chapters 14 and 15 for a quick review of this fitting procedure.) Figure 18-13 shows a rather low-quality fitting of old cambric around a square leg. A neater job is shown in FIG. 18-14, but a diagonal cut would have made it even better.

18-13 An old cambric featuring a rather low-quality fit at legs.

18-14 A little better fit, but you can see that no diagonal cut was made to give a neat finished edge around leg.

4. Fold edges *under* and staple close to the folded edge to create a tight, flat edge that won't be visible from a normal sitting position. Double-fold corners rather than take the time to cut out excess.

5. When fitting to curved areas (FIG. 18-15), trim the cambric to the outline, leaving a ½- to ¾-inch tuck-under allowance on all sides. Tuck and staple

18-15 Trimming cambric oversize to match curved perimeters.

the centers of the four opposing positions to get the cambric located (FIG. 18-16). Then work around the contour to create the finished edge all around. Often, with the cambric installed, it is difficult to see or locate screw holes for leg brackets, or t-nuts for leg bolts. Use a stuffing regulator (ice pick) to probe for holes.

6. When holes are located, cut an X in the fabric where bolts or screws must pass through. That reduces the tendency for the fabric to wrap around the screw or bolt as it is secured.

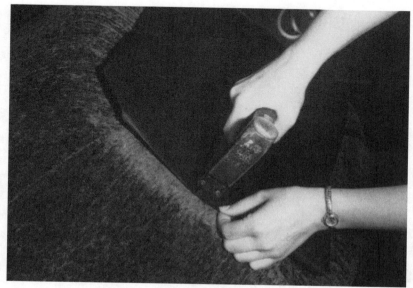

18-16 Even on curved furniture, begin stapling cambric at center of straighter areas and work toward corners.

Appendix

The checksheet recommended here is just that, a recommendation, not an absolute necessity or the last word in checksheet efficiency. It has been successfully used by numerous adult upholstery classes at Brigham Young University and seems to work well. However, if this checksheet does not seem to fit your requirements, feel free to make one that works better for you. But it is *strongly* recommended that some form of checksheet be used, especially by inexperienced upholsterers, so the unit can be put back together with success and pride.

INSTRUCTIONS FOR THIS CHECKSHEET:

Comfort test Sit in the unit and see how it feels to you. Is the seat too long, too firm, too soft, too low, too high? Is the back too short, too firm? How about the arms? Are they too low for comfort? Need more padding? Should they be softer?

Style changes Where and what widths of bands do you want? Do you want to add or remove panels? Do you wish to add or remove or restyle a skirt? Would it be better to change the methods of attaching certain components? Describe which ones and what the change is to be.

Cushions Check what kind they were. Then, in Notes area, describe what you wish to do when recovering the unit.

Cover List the pattern and color to be put back on the unit. Indicate the pattern repeat dimension.

Yardage chart List what the chart calls for.

Actually used Indicate how much was really needed or used.

EXACT ORDER OF FABRIC REMOVAL

This is the most important part of the checksheet. By recording with extreme care and accuracy which panels were removed first and how they were attached, a great deal of difficulty in determining what is to go back on first, second, etc. can be avoided.

Warning: When taking the various panels off, everything looks very simple and straightforward. But by the time you get to putting it all back together, it is surprising how much has been forgotten and how strange it all looks. Don't take shortcuts until significant experience has been gained.

Making sketches of areas and features that seem totally unfamiliar is also a great help, especially where several separate panels seem to (or actually do) come together at the same point. Every help given at this point will be doubly paid back at the time of re-covering.

Stripping Checksheet

1. Comfort test—Changes that are desired

    Seat: Depth_____" Firmness:

    Back: Height_____" Firmness:

    Arms: Height_____" Padding:

2. Style changes:

    Bands:

    Panels:

    Skirt:

    Methods of attaching:

    Cushions: Removable_____ Box_____ Knife edge_____ Waterfall_____ Attached_____

3. Notes:

    Cover: Pattern & Color_____ Repeat_____"

    Yardage: Chart_____ yds. Actually used:_____ yds.

Exact Order of Fabric Removal

	Part	Tacked	Tacking strip	Tack strip	Sewn	Blind stitched
Ex. 1	IB	Bot, T, R, L	NA	NA	T, R, L. (11 pcs)	T, corners
Ex. 2	OB	Bot	T, flex	R, L. Side	NA	NA
1.						
2.						
3.						
4.						
5.						
6.						
7.						
8.						
9.						
10.						
11.						
12.						
13.						

Separate sketches of different areas are a great help!

Comments:

Glossary

The definitions in this glossary have been written for descriptive and functional value. In some cases, uses, techniques, and other terms for reference have been added to enhance clarity and utility. Words appearing in italics will be found as separate entries in the Glossary. With some entries, the same term is used for more than one meaning. In those cases the separate definitions are numbered as (1), (2), etc.

accordion pleat   See *pleat: accordion.*

attached cushion   See *cushion: attached.*

auger bit   A drill used for making holes in wood. It has a tapered lead screw at the cutting end, an auger section for removing the wood chips, a round shank, and a tapered square end which fits into a grooved, two-jawed chuck of a brace.

Baker clip   A trade name for a 3- or 5-prong steel clip used to fasten edge wire to spring wires. The clip is applied with a spring-clip plier (also known by its trade name Baker clip plier).

balloon arm   An arm on a chair or couch stuffed so that it appears extra plump and full.

balloon cushion   A cushion of unusual thickness (greater than 5").

band   A narrow padded strip of cover material used to add "character" to certain furniture styles. May be located at lower portions of the seat, arm, and back and at the upper and top sections of arms, wings, and back rails.

band clamp   A simple clamping device made from a long piece of nylon strap that is tensioned with a ratchet-locked friction roller. Used to apply pressure around the full perimeter of articles to check fit and while glue is curing or drying.

bar clamp   A steel clamping device made from sections of pipe (½" to ¾") or band iron of varying lengths according to need. One end of the bar is fitted with a screw-thread movable jaw. The stationary jaw is adjustable along the length of the bar by notches in the band iron or self-locking, spring-loaded jaws on pipe sections.

beading   See *welt cord*, *welt*, and *welting.*

bench cushion   A single wide seat cushion that gives a unit the appearance of a bench on which two or more people can be seated comfortably.

bench seat   A seat wider than a chair, constructed as one continuous surface. It may have seams or welts but not separate or separate-appearing cushions, thus giving the appearance of a bench.

biscuit tufting   A style of tufting arranged in a square or rectangular pattern. See also *diamond tufting* and *tufting.*

blind stitch   A hand-sewn square stitch started from the underside of the cover fabric and then alternating from one side of the seam to the other. When properly done, none of the sewing thread is visible; hence the name.

blind tacking   A process of tacking a fabric to the frame from the back side so that no tacks or staples are exposed to view. This is accomplished through the use of a *tack strip, flexible tack strip*, or *tacking strip.*

blind tufting clip A metal clip used with the *blind tufting needle* to install covered buttons from the outside only. The clip, with button-tying twine attached, is forced through the cover and subsequent stuffing but not through the back cover. The blind tufting needle is then withdrawn, leaving the clip and twine inside the unit. The looped button is then tied on from the outside.

blind tufting needle A long (8" to 12") straight needle with a triangular point and a small hook a short distance from the point onto which a blind tufting clip is attached. Used to attach covered buttons to finished furniture where access to the inside for normal tying is restricted.

board foot See *measurement systems: board foot.*

box cushion See *cushion: box.*

box pleat See *pleat: box.*

boxing Side, front, and occasionally back panels of a box cushion.

brace (1) A hand tool having a two-jawed chuck at one end, an offset handle near the center, and a swivel pad at the other end, used to drive square-shanked tools such as auger bits. The combination is commonly called brace and bit. (2) See *brace block.*

brace and bit See *brace.*

brace block Short triangular wood pieces applied at the joints of uprights, posts, stumps, rails, and slats. The use of a brace block greatly increases the strength and rigidity of a frame.

bracket A steel-plate reinforcement used to provide added strength and rigidity to frame joints.

broad-faced seat band An unusually wide (more than 6 inches) panel of fabric which covers the entire front frame on a *unit.*

burlap (1) A coarsely woven jute cloth used to cover springs and open frame areas to serve as a support for stuffing and padding materials. It is easy to work and resists stretching and tearing. (2) Any wear-resistant, nonstretching fabric (other than jute composition) used to cover springs or open frame areas over which padding is to be placed.

button-making machine A hand tool designed for making covered buttons. There are a couple of different styles that are popular with upholsterers.

buttoned-pillow arm An arm of a chair or couch that is finished off on the top surface with a pillow cushion that also has been buttoned to the unit.

buttoning The process of applying covered buttons to a piece of furniture for decorative or contouring purposes. Similar to *tying.*

button-tufting needle A long (6-inch to 10-inch) straight needle having an eye 1 inch to 2 inches back from one end and a wooden knob at the other end, used for installing covered buttons that use an eye for attachment.

C-clamp A steel clamp shaped like a C, having an adjusting screw through one side that creates the clamping action.

cambric A black, slightly stiff, sized and calendered cotton or synthetic fabric used to cover the bottoms of seats to prevent dust and stuffing particles from falling to the floor. It comes in light and heavy weights and 30- or 36-inch widths. A white cambric is used for cushion and pillow covers. The original "cambric" was made of flax linen at Cambrae, France, whence the name.

camel back A couch or love seat that has one or two compound curves contouring the upper lines of the back.

channel A linear form of resilient padding, usually on inside backs and arms, around which the cover is usually sewn to a backing fabric making a pocket. A channel can be wide and relatively flat, fan shaped, curved and tapered, or narrow and rounded (pipe). Sometimes called *flute.*

chuck A component part of a tool which has two, three, or four opposing jaws for holding auger bits, twist drills, and other round, square, or hexagonal shanked tools.

claw tool A tool with a notched chisellike blade and an offset handle used primarily to remove tacks in the stripping process. The notch is useful in gripping the tack, but can become a nuisance because tacks may lodge in the slot or it requires rather close alignment. Sometimes called (though not preferred) *tack remover*.

coil spring See *spring: coil.*

comfort factor Term used to identify the comfort of a foam. It is determined between the IFD reading of 25 percent and the sag reading at 65 percent deflection. Special foams, (HR) can reach a sag factor as high as 2.4. No conventional foam can achieve that level of performance.

cording See *welt, welt cord, welting.*

cotton A natural fiber or fabric made from the cotton plant. As a fabric, it is not popular for a cover. See *cotton felt.*

cotton felt A 1-inch thick (approximately) mat of loosely padded cotton fibers composed of varying percentages of gin flues, linters, staple and first-cut fibers; also comes in cotton-polyester combinations as well as flame-retardant varieties. Used for stuffing and padding and is sold in rolls 27 inches wide and approximately 20 pounds per roll. Most commonly and simply referred to as *cotton.*

couch A furniture piece designed to seat three or more people. Usually contains three or more cushions or panels, although some styles call for a two-cushion or panel unit. It is usually 72 inches or more in length and is frequently referred to as a *sofa.*

cover The outer fabric of upholstered furniture.

covered button A button covered with an upholstery fabric. The back of the button may have a nail, split prong, or a loop by which it is attached to the unit.

cross tying A method of tying coil springs for the purpose of providing a more firm, even spring support (see also *eight-point tie, four-point tie*).

crown The line defined by a 45-degree intersection of the horizontal top surface and the vertical side surface of an arm, back, or seat. It is the line at which cuts must be made in cover fabric for proper fitting.

crowned back A couch or loveseat having a back with a single, smooth upward curve that extends from one side to the other.

Curve-Ease A brand name for flexible tack strip.

cushion: attached Any cushion (back or seat) that is affixed to the main furniture piece by sewing, buttoning, or tying.

cushion: box A cushion having side panels and principally square corners. This gives the cushion a boxlike appearance. May be made with or without welts.

cushion: clam-shell A cushion (usually back) that has an extra-wide, extra-deep single tuck sewn into it with the entire length pulled into the stuffing so that multiple wrinkles radiate from the tuck, giving the appearance of a closed clam shell.

cushion: fixed See *cushion: attached.*

cushion: knife-edge Cushion with seams, and usually large welts, located at the front and occasionally side centerlines of seat cushions or the top and occasionally side centerlines of back cushions. The knife edge gives the cushion sides a rounded appearance. Welts are usually made with ³⁄₁₆-inch to ⅜-inch diameter cord, depending on size and style desired. (See also *full knife-edge.*)

cushion: L or J A cushion which makes a short right-angle bend in front of the arm stump. A cushion passing in front of the right stump is the L cushion; the one passing in front of the left stump is the J cushion. These cushions are usually found on love seats, couches, or sectional units.

cushion: loose Any cushion (back or seat) that is not fastened to the main furniture piece.

cushion: ram-horn A cushion (usually back) that has the top stuffing tapered and the edges of the cover rolled under the ends so that the side view resembles a ram's horn.

cushion retaining groove A recessed linear region created 2 inches to 5 inches back (depending on the style) from the front edge of the deck that acts as a retainer for a loose cushion reducing the tendency to creep out of the seat and permits the cushion to nestle in by reducing the gap between the somewhat rounded cushion face and the deck.

cushion size The outermost dimensions of a stuffed, completed cushion, including the side bulging that occurs due to the oversized stuffing. This size is usually considered to be ½ inch greater per direction than the finish size.

cushion: squared-crescent A pillow cushion having an inner contour in the shape of a crescent and the outer contour of virtually square corners.

cushion: T A chair cushion that has a short section that goes in front of both arm stumps.

cushion: waterfall A cushion on which the fabric continues in one piece from the back, around the front (waterfalling), and terminates at the rear of the opposing face. The boxing on this style is rounded at the front and is usually squared at the back. May be constructed with or without welts.

cut size The size of a piece of fabric that includes all tucks, pleats, and seam allowances added to the finish sizes.

Dacron (1) A curly white polyester fiber mat used as an outer layer or built-up layers of padding to provide extra softness in the final feel. It is also used to stuff pillows. Sold in rolls 27 inches wide weighing nominally between 6 to 7 pounds. (2) A brand name for a thermoplastic polyester fiber or fabric.

davenport See *couch* and *sofa*.

deck The base area upon which the seat cushion (fixed or loose) rests.

decorative tacks Tacks or nails with large domed heads that may have a hammered, antiqued, polished, brushed, geometrically contoured or combination of these finishes. They are made of brass, aluminum, steel, or stainless steel; they are plated, painted, or plastic coated. Used to add decoration, characterize period furniture, or strike a special motif.

decking A durable fabric, cotton or synthetic, used to cover that part of the deck not normally visible when cushions are in place. It is often used to reduce cost, since it is significantly less expensive than most cover materials.

density The "weight" of a unit volume of foam, expressed in pounds per cubic foot. Many people erroneously use this term to indicate "hardness." See *IFD* for comparison. Most foams will have densities up to 2 pounds per cubit foot with seating foams being in the range of 1.4 to 2.0 pounds per cubit foot (pcf).

diagonal cut A single straight-line cut made in a piece of cover, muslin, or burlap that goes diagonally toward the inside corner of a post, stump, or rail where that fabric is to be fitted to two sides only of the frame member.

diagonal cutters A plierlike tool having a pair of cutting jaws that are placed at a slight angle to the handles (diagonal thereto) used for cutting wires and cords and extracting staples and remnants.

diagonal tying See *eight-point tying*.

diamond tufting A tufting operation where the tufts have a diamond shape. See also *tufting*.

diamond tying See *eight-point tying*.

dikes The abbreviated popular name for *diagonal cutters*.

double-point needle A straight needle pointed at both ends. The points may be round (used for woven fabrics) or triangular (used for vinyls and leathers), having a single eye a short distance (between 1 inch and 2 inches) from one end. Used for sewing cushion retainer grooves or installing covered buttons when a button-tufting needle is not available.

dust cover   A fabric to prevent dust and stuffing materials from falling to the floor. See *cambric*.

edge roll   Generally a 1¼-inch diameter roll of twisted fiber-core material covered with 10-ounce burlap or polyester fabric and sewn snugly to form. Most edge rolls have some form of fastening lip and are fastened directly to the front seat or top back frame or to the edge wire. Most in use today are commercially produced. In years past, upholsterers formed, wrapped, and sewed their own. See also *frame edging*.

edge wire   Heavy steel wire, 8- or 9-gauge, used to form a straight edge on coil-spring construction or on sinuous-spring systems incorporating the V-arc and Z-arc edge suspension. Edge wire is attached with metal clips called Baker clips or spring clips.

eight-point tie   A method of tying coil springs using four pieces of typing twine tied at 45-degree angles to each other, with two ties occurring per strand on each coil. Also referred to as *cross tie*. See also *four-point tie* for comparison.

fabric saw   A power-driven (pneumatic or electric) saw used by production upholsterers, commercial textile fabricators, and tailors for cutting multiple layers of fabric to the same shapes and sizes. The saws will have counterreciprocating very fine-toothed saw blades or a smooth disk blade.

fatigue   The loss in load-bearing quality of a foam. The popular test for fatigue is derived from a static load being applied to an IFD of 25 percent for 17 hours at room temperature. The IFD loss is then expressed as a percentage of the original IFD value.

feathering   The act of tapering the edge of a stuffing or padding material to give a smooth-tapered look to the cover, leaving no end-of-padding lines.

finish size   The size of a piece of fabric or cushion from seam stitch to seam stitch. Seam allowances are added to these dimensions to obtain the cut size. Refer also to *cushion size*.

fitting   The process of stretching, cutting, and tacking the cover into final location.

flap panels   Thin, lightly padded cushions constructed on the order of a pillow, attached to the arm or back (or both) of a unit by a flap sewn to the back of the cushion near the top. If these cushions are not further attached through the use of covered buttons they can be flapped upward by the flick of the hand, thus the name.

flexible tack strip   A flexible, notched metal strip that is formed to a right angle of approximately ½ inch each side. Used extensively for fastening outside arm and back panels having curved lines.

flop cushion   A cushion resembling a pillow, completely enclosed, usually having a zipped back and attached to a piece of furniture with an extra flap located near the rear upper edge of the cushion permitting it to "flop."

flute   See *channel*.

foam   The abbreviated term most commonly used in upholstering practices to refer to *foam rubber*.

foam adhesive   An adhesive, usually in aerosol spray cans, formulated especially for attaching foam rubber materials to each other or to other fabrics. It is often used to fasten fabrics to other base materials such as wood or metal. This adhesive, when cured, remains flexible, permitting resiliency while maintaining adhesion.

foam rubber   A porous, usually open-celled, urethane-based flexible material. Other materials have been used in the past but with limited success. The industry uses urethane foams almost entirely. Most frequently it is referred to as just *foam*.

foam saw   A saw especially designed to cut slabbed foam to smaller or contoured shapes. It is usually an electric handsaw with two counterreciprocating fine-toothed blades and a tapered base.

foundation   The spring units, webbing, or solid (wood, plastic, or metal) base that provides the general support over which desired stuffing and padding is placed and gives the desired resiliency or rigidity for the furniture piece.

four-point stay The application of tacks or staples at four points, usually on an inside arm panel, to hold it in place while final fitting is completed.

four-point tie The two-directional method of tying coil springs, involving two pieces of tying twine with two ties occurring per strand on each coil. This is the minimum tie usually used and gives the softest suspension. The ties are made at 90 degrees to each other and square (as far as possible) with the frame. Refer to *eight-point tie* for comparison.

Fox edge A rolled edging used to soften, round, and in some cases extend rail, post, and stump edges. Fox edging is a brand name and should not be mistaken for edge roll but fits the frame-edging classification.

frame edging A roll of fiber or plastic material ranging from ⅜ to 1 inch in diameter, jute or polyester covered, with or without a fastening lip, designed to give sharp edges a rounded, softened effect. It is used also to provide a recess for panels and reduce fabric wear at corners and edges. Refer to *edge roll* for comparison. Also called *Fox edge*.

full knife-edge A knife-edge cushion on which the center seam extends not only across the front (seat) or top (back) but around the sides.

gauge The diameter of wire (edge wire, etc.) and the thickness of sheet metal used in plates, brackets, and braces is expressed using this term. The higher the gauge number, the smaller the diameter of wire or thinner the sheet metal.

gimp A narrow band of decorative material used to conceal tacks, staples, and cover edges that would otherwise be visible and objectionable. It comes in ribbon form and is purchased by the yard, card, or roll.

gimp gun A pneumatic staple gun that uses narrow-back staples especially suited for very low visibility in fastening gimp to upholstered units. It is also used for holding down other areas of cover where seeing the staple would be objectionable.

gimp tack A tack having a small-diameter domed head, used to attach gimp or hold down corners, tabs, pleats, or folds of cover material. The small head virtually disappears from view when used properly.

glide Metallic or plastic (usually nylon) domes or caps placed on the bottoms of legs to reduce friction, marring, and snagging of floor surfaces.

hand screw A clamping device having two counterthreaded steel rods with handles and two wooden jaws that can be adjusted to apply pressure to parallel or slightly non-parallel surfaces. Used primarily to hold wooden pieces together while glue sets or dries.

hard edge A narrow strip of wood, often plywood, used as a retainer for the seat-cover panel and as an elevator to raise the front seat rail to a height slightly above the crowned height of the springs.

helical spring See *spring: helical.*

high-resiliency foam A urethane foam (classified HR) possessing an exceptionally high degree of "life," or resiliency. Although more expensive than conventional foams, HR significantly increases the comfort factor and durability beyond that expected from the conventional urethane and latex foams.

hog ring A steel wire formed in the shape of a C having pointed ends and measuring approximately ¾ inch across its width. It is used extensively in auto upholstering and in areas where fabric is to be fastened to wire or rod anchoring systems. The rings are most successfully applied with *hog-ring pliers.*

hog-ring pliers A plier having jaws with recessed grooves, especially designed for holding hog rings. There are standard and spring-loaded models, the latter being much more convenient to use as they retain the ring within the plier without the need of any further pressure being applied by the operator. There are also straight and angled models.

HR See *high resiliency.*

IFD Indentation force deflection: a measure of load bearing (hardness of a foam material). A specimen $15 \times 15 \times 4$ inches is placed under a circular plate of a 50-square-inch area, which is depressed 1 inch. This is 25 percent indentation. A reading is taken that shows the number of pounds required to make that deflection. If 20 pounds were required, the foam IFD would be 20 pounds at 25 percent deflection.

In line The abnormal but occasionally necessary practice of applying staples parallel, or nearly so, to a fold or tuck rather than perpendicular to it. This practice most often occurs when arming round-topped padded arms at the stump.

interlacing Crossing of one webbing strip alternately over and under other strips, giving a uniform support base for coil springs or other stuffing.

innerspring Coil springs contained within individual fabric pockets or several coils linked together and enclosed as a group in a fabric (*Marshall unit*).

jute A very durable imported (from India) natural plant fiber that resists tearing and stretching. It is used extensively in burlap, webbing, and twines.

Klinch-it clip Metal clips used to anchor coil springs to cloth webbing. The clips are applied with a Klinch-it tool, which presses the clip prongs through the webbing and spreads them sideways, clinching the spring tightly to the webbing.

Klinch-it tool A dispensing-type tool having a long throat or tube into which Klinch-it clips are placed, used for anchoring coil springs to webbing with the specially designed clips (*Klinch-it clips*).

knife edge See *cushion: knife edge*.

lead screw A short tapered screwlike thread at the point of an auger bit that draws the cutting edges into the wood (until the lead screw extends out the back side of the piece being drilled).

legs The two parallel sides of a staple that penetrate into the substrate.

loose cushion See *cushion: loose*.

love seat A furniture piece designed to seat two people comfortably. It is wider than a chair and narrower than a couch, the normal length being from 54 to 66 inches. Can be made with a one-piece seat cushion or with two sections or cushions, the latter being the most popular.

low-profile spring See *spring: sinuous*.

mallet A rubber, wooden, plastic, or rawhide tool used to drive ripping tools or claw tools, set tack strip, and smooth stuffing and padding after cover is in place. Rawhide or white rubber are preferred by most upholsterers.

Marshall unit Coil springs sewn in individual muslin or burlap pockets and fastened together in strips or complete ready-made units; used in seats, backs, cushions, and mattresses. Also called *innerspring* units.

matching band A band of the same proportional width as the remainder of the surface to which it is attached, usually along the front seat frame.

measurement systems:

  board foot A surface area of any material which is 1 foot on each side (1 square foot) and 1 inch thick. It is calculated by multiplying the length of the material by the width (L $\times$ W) either in inches or feet. If calculated in inches, conversion to board feet is made by (1) dividing the product by 144 (the number of square inches in 1 square foot) and (2) multiplying this product by the number of inches the material is thick. If calculated in feet, the area is already in terms of square feet. Merely multiply the product by the number of inches the material is thick.

  linear Derived from *line*, this measurement pertains to one direction only—length. Examples of materials for which the linear measure is used: ribbon, gimp, rope, twine, wire and wood moldings. Fabric, cotton, Dacron, and some other materials are often sold by linear measure. Also referred to as *running measure*.

metric Although much has been done to prepare the United States for a conversion from the English measuring system to metrics, almost no binding activity has been noted in the area of upholstery and textiles yet. Even the spelling of the standard for linear measure, the *meter* (as it is preferred by the U.S. Metric Board) is often spelled *metre*, which is the preferred spelling of the ISO (International Standards Office).

running A measurement referring to length only (see *linear*).

measuring The process of taking measurements with a cloth or metal tape from a finished piece of furniture to establish the proper cushion or panel size. Refer to *tailoring* for comparison. Measuring for establishing cushion dimensions works well with straight-line seats. (*Tailoring* is recommended for units with any curvature in seat.)

metalene nails Nails having extra-large flat heads that have been painted or vinyl coated to blend in with vinyl coverings. They are used to apply gimp and in some instances as decorative nails.

muslin A lightweight, inexpensive cotton cloth used as a first cover over the completely padded furniture. Especially valuable for the learner since it provides experience in tailoring or measuring, cutting, and fitting operations before the much more expensive cover material is worked.

nail head, set trim A finishing style wherein the IB or IA panels are tacked to the front and tops of the frame, respectively, then a band panel is attached with decorative tacks and stuffed to created a rounded band.

narrow band A band added to a piece of furniture that is proportionately more narrow than the remainder of the surface to which it is attached.

needle-nose A plier-like tool having long tapered jaws, used to bend small radii in soft wires, grasp small objects in restricted areas, hold gimp tacks for starting, and similar uses.

No-sag spring crimper Part of a hand-tool unit, originally designed by the No-Sag Spring Company, used for reversing the bend in the cut sinuous spring.

No-sag cutter Part of a hand-tool unit, originally designed by the No-Sag Spring Company, having a single hardened-steel cutting blade that is pressed toward but stops at the face of a steel cylinder around which a section of sinuous spring is placed and subsequently cut.

no-sag spring See *spring: sinuous*.

notches Small cuts made in the seam-allowance area of cover panels for the purpose of providing matching and aligning marks in the pieces to be sewn together. The cuts are usually made in a V shape but may also be cut as double and triple V's; single, double, or triple U's; combinations of V's and U's; or simply as slits.

overstuffed A piece of furniture on which virtually the entire surface is covered with a cloth or vinyl fabric except for relatively small sections of show wood. Implied in this classification is generally the notion that the unit has one or more types of spring foundation and padding.

padding The outer layer of resilient material just beneath the cover giving the desired feel (firmness) or appearance (plumpness) to the furniture unit. It may be used over the top of stuffing materials or alone, in which case it is both stuffing and padding, as in super-soft foam for a back cushion.

panel (1) A piece of heavy cardboard, light (⅛ to ¼ inch) plywood, or similar material cut to the matching contour of the front of an arm stump, the ends of a back, or the outside of a wing. The panel is added (usually), covered, and nailed in place (usually with 4-penny finish nails or wire brads), giving the stump, back or wing a pleasing finished look. (2) Any of the major pieces of cover fabric cut and readied for application, such as seat panel, arm panel, back panel, zipper panel.

pattern repeat Refers to the distance from one point on a fabric to the nearest point at which the pattern repeats itself. Ranges go from 3 to 27 inches generally, only a few being less or greater than this range.

pattern layout A freehand sketch of all the panels of fabric that is needed to cover or recover any given unit.

Phillips screwdriver A screwdriver having a tapered, four-waned blade that fits the recessed Phillips-head screws.

pillow arms An arm of a chair or couch that has as a final padding an attached pillow cushion.

pillow back The back of a chair or sofa that has a pillow cushion as the final padding.

pillow cushion A cushion that is constructed much like a pillow, enclosed on all sides and having no boxing.

pillow spring See *spring: pillow*.

pipe A channel produced by filling rather narrow tubes of cover material that are sewed to a fabric backing, giving a pronounced rounded appearance similar to the pipes of a pipe organ, whence the name. Generally pipes are more slender than channels, although the terms are occasionally used interchangeably.

place-tacking See *stay-tacking*.

platform (1)The wooden or metal structure upon which a piece of furniture rests, having solid footing on the floor, permitting the rest of the item to rock, swivel, or both. (2) The flat, horizontal portion of a solid seat before installation of cushions, stuffing, or padding.

pleat Regular folds sewn into the cover fabric, usually associated with skirts, tufts, channels, or pipes. Also used at corners and other areas requiring a reduction of the perimeter of the fabric. Refer to *tuck* for comparison.

pleat: accordion A pleat made with the folds all going the same direction, giving a stepped effect.

pleat: box A pleat made by alternating the direction of the folds, giving an alternating planar effect.

pleat: kick A small section of cover made in same manner as the rest of the skirt but separate from it and attached to the frame at corners or at leg positions to provide complete coverage, giving the rest of the skirt mobility at these locations.

Pli-Grip A brand name for a notched, metal flexible tack strip capable of forming compound curves with ease.

polyfoam An abbreviated term referring to flexible urethane foam rubber of a "standard" firmness (IFD) and resiliency.

post The front vertical member of an arm or back corners; normally narrower than 3 inches; arm may be covered, partially covered, or all show wood; does not involve use of a panel (compare with *stump*).

nail The horizontal members of a frame that give the furniture the general form and structure.

recovery The "bounce-back" of foam material. It is derived after the sag factor has been taken. The foam is again depressed to 25 percent IFD. This second reading is divided by the first 25 percent IFD reading to give a percentage of recovery. High recovery factors are essential for good cushioning applications.

regulator See *stuffing regulator*.

right angle An angle of 90 degrees; one that makes a square corner.

ripping tool A chisellike tool with a sharpened, square-ended, offset blade, used primarily for removing tacks in the stripping process. Refer to *claw tool* for comparison.

rubberized hair Curled animal hair (generally hog) coated with a rubber film and processed in mat form approximately 1 inch thick. Used as a base stuffing in earlier upholstering techniques. Although still available, many upholsterers no longer use it; it gives a rather stiff, "crackling" base and mats down rather readily.

running stitch A straight stitch made close to and parallel with the seam, giving a very visible stitched appearance. It may be made on either or both sides of the seam. In some vinyl upholstery, a double running stitch is used to attach a fabric reinforcement strip to the back of the seam to relieve the stress on the vinyl seam and reduce potential tearout. See also *top stitching*.

sag factor The cushioning quality or resistance of a foam to "bottoming out," derived from the ratio of 65 percent IFD to 25 percent IFD. A sag factor of 1.8 to 2.0 is considered normal and is referred to as "standard."

sagless spring See *spring: sinuous*.

scissor stapler A hand-held stapler having handles at one end, a pivot in the center, and the stapling end at the opposite end. It is used extensively in professional upholstery to hold folds and tucks in place temporarily for sewing. It also finds frequent use in fastening Dacron edges together around cushion foam or spring units in preparation for stuffing into the cushion cover.

screwdriver bit A screwdriver tool having a point (straight-slot, Phillips, etc.) and shank but no handle. The shank end is round, square, or with "ears" for inserting into a Jacob's chuck or brace, respectively.

seam allowance The amount of fabric extending beyond a seam that is not a visible part of the cover panel. For consistency and ease of reference, ½ inch has been designated for use in this text.

seat band See *band*.

sectional A set of two or more furniture pieces that can be used singly or in combination to form a variety of seating arrangements. These pieces are designed so that they can be put next to each other and look almost as if they were one piece.

show wood The finished wood surfaces that are supposed to be exposed when the upholstery is complete.

silencer A strip of fabric, oil-impregnated craft paper, webbing, foam rubber, or other stuffing material used to reduce or eliminate noise of springs moving against the frame clips or support members.

sinuous spring See *spring: sinuous*.

skewer A slender pin, usually 3 to 3½ inches long, having a sharp round tapered point at one end and a ring "handle" at the other. Also known as "upholsterer's pin."

skirt A horizontal segment of fabric, usually around the bottom of a piece of furniture that is fastened (blind-tacked or sewn) at the upper edge only, leaving the bottom portion free to move. May be straight or pleated. Refer to *pleat: accordion, pleat: box*, and *pleat: kick*.

slat A horizontal frame member between rails or arms that adds strength and support and provides for tacking and stretching the cover. Refer also to *tack rails*.

slip-joint pliers The proper term for common household pliers. It is thus called because the pivot joint "slips" at the full-open position to accommodate size adjustments.

slip seat An upholstered frame seat constructed for easy removal and installation to a furniture frame. Usually attached by means of wood screws or bolts.

slip-tacking See *stay-tacking*.

socks Muslin casings for individual and rows of coil springs from which Marshall units are made.

sofa Another term for *couch*. There is no universal distinction between *sofa, couch*, and *davenport*. All three terms generally refer to a unit long enough to accommodate an adult reclining.

spring The major suspension component for furniture provided by spring-steel coils, bends, or curves.

spring-bar unit A group of two to five single-tapered coil springs fastened to a steel support bar that is fastened to the frame of the furniture. Generally used for seat suspension.

spring clamp A clamp that operates much like a pair of pliers except that a spring holds the handles apart and the jaws together.

spring clip The generic term for steel clips used to fasten two wire segments together. Refer also to *Baker clip*.

spring-clip plier A special plier having offset pronged jaws designed for crimping the tabs of spring clips around edge wire and spring elements.

spring: coil Coils of spring-steel wire used as a major resilient base in furniture and mattresses. Coils can be wound with parallel sides or with single or double tapers. In double tapers, the smaller diameter is in the center of the spring rather than at the ends.

spring edging An edge roll made in varying sizes to close the gap between the deck edge and the cushion. It also helps keep cushions in place while reducing the sharpness of the spring edge.

spring: helical Small diameter (½ to ¾ inch) coils of spring wire wound parallel, used (occasionally) as a connector between sinuous springs, between edge wires and springs, and (frequently) between springs or edge wires and the frame. Lightweight coils are used for backs, heavier weight for the seats. They come in open and closed form.

spring: low profile See *spring: sinuous*.

spring: No-Sag A brand name for a sinuous spring manufactured by No-Sag Spring Division of Lear Siegler, Inc. No-Sag has become the catchword in the industry, probably because the company was the first on the market with the product; hence it carries the company name, much like Formica, Styrofoam, and Masonite—all brand names for products. Refer to *spring: sinuous* for gauge size and use area.

spring: pillow A lightweight, basically parallel wound coil spring with both ends fastened with a clip to the last coil, preventing rips or punctures to a fabric. In spite of its name, this spring is not used in pillows.

spring: sagless See *spring: sinuous*.

spring: sinuous A steel-spring wire bent in a serpentine form and wound on a roll. Also called *No-Sag, zigzag*, and *sagless spring*. Heavy-gauge wire (8 or 9) is used for seats, light-gauge (1 or 12) is for backs.

spring: zig-zag See *spring: sinuous*.

staple A wire fastener having a square U shape used for tacking upholstery materials to wooden or structural foam (plastic) frame members.

staple gun A pneumatic, electric, or manually operated tool that drives wire staples at the press of a trigger. It is estimated that the use of a staple gun reduces upholstery work time 30 to 50 percent from the earlier use of tacks and tack hammer.

staple remover A tool used primarily for removing staples. There are two popular varieties: a flat, two-pronged, patented model (commonly called the *Berry picker*, after the name of the inventor) and a flat, wedge-shaped tool. Each has its advantage, but the Berry picker is preferred by most upholsterers and is discussed exclusively in this text except for the tool identification section where the wedge tool is illustrated.

stay-tacking The procedure of temporarily fastening a cover panel into basic position by driving a number 8 or larger upholstery tack partway in to hold the panel from shifting while final fitting is pursued. Stay tacks are easily removed when desired. Also called *place-tacking* or *slip-tacking*. Staples are often used in modern practice instead of tacks.

stretcher A strip of scrap material (cover, burlap, decking, etc.) sewn to the cover in areas where additional length is needed to stretch and tack and where the stretcher will not be seen when the unit is in normal use. The use of a stretcher can save on yardage of cover required, depending on size and cutting orientation.

stripping The process of removing old fabric and damaged padding, stuffing, springs, or other support materials.

stuffing The underlayers of resilient materials used to provide basic softness or firmness to the upholstered item. All upholstered furniture uses some form of stuffing. *Padding* and *stuffing* are often used interchangeably.

stuffing regulator A sharply pointed, tapered steel instrument used for inserting through cover or muslin cover to move (regulate) small amounts of padding without having to remove the panels.

stuffing tool A tapered, wedge-shaped tool used to force stuffing or padding into corners, channels, pipes, and areas that cannot be reached by hand.

stump The front vertical member of an arm. A stump is usually wider than the average arm post, being from 3 to 8 inches wide, may be contoured, is most often finished with a panel, and involves a covered arm.

supersoft foam A very soft foam rubber used to make very soft cushions or padding (usually restricted to backs). It has just enough resiliency to return the fabric to its desired shape without giving a firm feel (usually about 10 IFD).

tab (1) The V-shaped center portion of the Y cut that is tucked under to create a "finished edge" when the cover is fitted against a post or upright. (2) A strip of cover fabric that is cut in a panel (like the rear portion of an inside arm panel) to permit stretching of the fabric on either side of a rail or around the crown.

tack (1) A metal fastening device characterized by a tapering shank (except in decorative tacks) and a variety of head styles. (2) A term used to signify attaching a fabric to the frame of the furniture. The term originated from the earlier exclusive use of tacks. Modern practices use staples for the same purpose; hence, tacking a fabric panel in place may be done with staples.

tack hammer A curved, slim, double-headed hammer having one head magnetized to pick up and hold tacks, making one-hand installation possible. The flat sides of the hammer are often used to install fiber or metal tack strip, thus preventing cutting the fabric. Also referred to as *upholsterer's hammer.*

tacking The process of fastening fabric materials to the frame. Until the early 1950s, upholstery tacks were used exclusively for this task, hence the term. Even though staples are used by most shops today, the term *tacking* is still used.

tacking strip A chipboard strip purchased in rolls or strips (sometimes in sheets, then cut into strips) used for blind-tacking the cover where straight sections can be reached from the underside of the fabric, like the tops of outside arm and back panels. Most common strip width is ½ inch; also comes in a lightweight, easy-to-bend metal strip (refer to *flexible tack strip*).

tackless (1) A term used by some upholsterers to refer to a tack strip because, like its namesake, it requires no addition of tacks. (2) The term is more correctly used to refer to the metal or wooden strips used for fastening carpeting to the substrates without the necessity of additional tacks.

tack rails A horizontal frame member, especially installed to provide a tacking surface in locations where the rails or slats do not meet upholstering needs.

tack remover See *claw tool, ripping tool.*

tack strip A rigid chipboard or metal strip into which tacks have been affixed by the manufacturer, used to accomplish straight blind-tacking, which must be installed and finished from the outside. Comes in strips 27 or 30 inches long with tack sizes of 8, 10, or 12 ounces.

tailoring The process of laying a cushion panel on a seat (or back) of a finished piece of furniture and marking with chalk to the exact contour of the seat or back area, thus tailoring the panel to the precise size and shape. This process is more accurate and a little more time-consuming than measuring.

tear-out Fabric tearing away from a stapled or tacked area because of loose or insufficient fastening, excessive pressure, or failure to tack an externally folded flap over the top of a partially attached panel.

top stitching The process of sewing a normal seam again on one (single top stitch) or both (double top stitch) sides of the original seam, creating a styling change as well as a stronger seam. The increased strength is accomplished by folding the seam flaps to one side and sewing through all layers from the top side of the fabric. The styling change is created by the thread of the top stitches, which are visible from the face side of the fabric. The thread may be in a contrasting or blending color. Additional reinforcement may be added by sewing a narrow strip of fabric to the back side of the seam. This is restricted to the double top stitch, however. Sometimes called a *running stitch*.

trestle   A sawhorse-type stand with a padded, covered top used for elevating and supporting furniture during the upholstering process. The padded top, which is also recessed to retain furniture legs, protects furniture from being scratched and fabric from being snagged or torn.

tuck   A fold made in cover fabric to reduce the perimeter.

tufting   The process of using covered buttons, pleats, tucks, and tying twine in a patterned array to hold the cover and padding in place, giving styling and contouring to the furniture. Characterized by raised areas (tufts) that are generally diamond (*diamond tufting*) or rectangular (*biscuit tufting*) in shape.

twist drill   A drill having a smooth, round shank, usually two flutes, and a smooth tapered point with no lead screw. The flutes look as if a flat piece of steel had been twisted. It is designed primarily for drilling metals and other materials of equal or less hardness.

tying   The process of tying a strong twine to the seam flaps at intersections or through the cover fabric and padding to create depressions for design or contouring or affixing cushions to the furniture.

tying twine   A tough, heavy twine, tightly twisted or braided of jute, nylon, polyester, flax or polypropylene used for tying springs, buttons, and edge wires. Comes in 1-, 2-, and 10-pound spools.

unit   (1) A complete piece of furniture such as a chair, ottoman, couch, recliner. (2) A set or subassembly of pieces that can be removed as an entity, such as a spring unit, back or arm assembly, footrest, cushion.

upholstered   Furniture with some form of padding, stuffing, springs, or combination thereof covered by cloth or vinyl fabric.

upholsterer's hammer   See *tack hammer*.

upholsterer's pin   See *skewer*.

upholsterer's shears   A heavy-duty pair of shears, improperly referred to by some as "scissors," used to cut heavier upholstery fabrics, twines, and threads.

upright   Vertical support members of a frame that do not serve as legs, posts, or stumps.

V arc   A sinuous spring construction, the end of the spring formed in a V shape. This makes it possible for the end of the spring to have a significant "spring" or return not normally found with the standard sinuous construction. If it is part of the sinuous spring, it requires an additional slat for fastening the spring end since the end is now at a position shorter than that of the seat or top rail where the sinuous spring is normally attached. If it is a separate section, it fastens to the edge wire and rail. See also *Z arc*.

vents   Holes made in solid seat boards when a seat is covered with vinyl or any nonbreathing cover fabric. If vents were not provided, the seat would become an air pillow.

webbing   A strip of jute, cotton, plastic, rubber, or metal ranging from 1 to 4 inches in width and of various weights (thicknesses), used to provide a foundation for coil springs and other stuffing materials. Plastic webbing is often used as the only webbing material on furniture like lawn furniture.

webbing pliers   A plier having corrugated, parallel jaws (usually 3½ inches wide) used to grip and stretch webbing, leather or other fabrics.

webbing stretcher   A tool used to grasp webbing by pressure, puncturing, or binding to stretch it taut over the furniture frame. The tool may be designed with jaws, tapered steel pins, or a slotted head.

webbing tack   A barbed tack used to fasten webbing to wooden or structural plastic frames. The barbs provide extra holding power, which is helpful because of the heavy stresses placed on the tacks holding webbing.

welt   (1) A strip of cover fabric that has been sewn around a welt cord making a round, decorative edge that can be sewn onto other cover panels or tacked directly to the frame. (2) A cord-filled edging applied to cushions (box, knife-edge, and waterfall).

welt board   A straight length of board measuring 1½ to 2½ inches wide. The thickness and length are optional, usually ¾ × 72 inches, made of alder, poplar, birch, maple, or similar close-grained hardwoods.

welt cord   A round cord (purchased in rolls) of varying diameters used as a core around which upholstery fabric is formed to make *welts*. Made of a foamed polyethylene (preferred by most upholsterers today), twisted jute, or wrapped paper core.

welting   Strips of cover fabric before they have been made into a *welt*. Usually cut 1½ inches wide and of the length necessary. Upholsterers often refer to these strips simply as "welt strips."

wrapped arm   An arm style; the inside arm panel wraps over the top of the arm and completely around the front of the post or stump. This style is distinguished by pronounced large, outward tapering tucks as the fabric rounds over from the front to the top of the arm.

Y cut   A straight cut made in the cover, muslin, or burlap that goes toward the center of a post, stump, or rail and terminates by making two angular cuts, each toward the corners of the frame member. It is called the Y cut because it looks like a Y cut into the fabric. This cut is used to provide a means for folding the fabric inward and fitting to three sides of a frame member, giving a finished edge. See also *diagonal cut*.

Z arc   A sinuous-spring construction, the end formed to the shape of a Z, or a commercially prepared spring unit of the same shape, used to give the spring edge a springiness similar to coil-spring construction. The Z arc provides a softer spring than the V arc. The end of a Z arc is fastened directly to the front seat or top back rail. See *V arc* for comparison.

zigzag spring   See *spring: sinuous*.

zipper glide   Metal or plastic unit that glides along the zipper stock, opening and closing the zipper.

zipper panel   A panel of cover fabric into which the zipper is sewn, usually in cushions. For the hidden zipper, the panel is cut 1 inch wider (assuming a ½-inch seam allowance) than the cushion boxing. For a centered zipper, the panel is cut lengthwise in half.

zipper pocket   A pocket made at one or both ends of the boxing to fold over and conceal the zipper glide when it is in the full-open or full-closed position. The pocket is made by making a Z or trifold in the boxing. Pocket depth varies, the normal 1 to 1½ inches.

zipper stock   A continuous strip of zipper material (purchased by the foot, yard, or spool) from which any length zipper can be cut. It comes in several sizes. Zipper glides are purchased separately to match the stock size.

Index

OTHER RELATED BESTSELLERS OF INTEREST

Refinishing Old Furniture

George Wagoner

A do-it-yourselfer's guide to restoring old and damaged furniture to it's original showroom condition, with detailed, step-by-step instructions, working photographs, and illustrations and important safety guidelines.
Paper 0-8306-3496-7 $13.95

Fixing Furniture: An All Thumbs Guide

Robert W. Wood

How to restore damaged, worn, or faded wood furniture at home—absolutely no woodworking experience required. Otabind, 2-color throughout.
Paper 0-8306-4433-4 $9.95

Painting, Wallpapering, and Stenciling

Robert W. Wood

This All Thumbs Guide is just the thing for homeowners who have always wanted to add decorative touches to their homes, but have hesitated for fear of making costly mistakes. Like the other book in the series, Detachable Tip Cards, lay-flat binding, and concise, illustrated instructions for even the most reluctant do-it-yourselfer. Two-color format.
Paper 0-8306-2547-X $9.95

Look for These and Other TAB Books at Your Local Bookstore

To Order Call Toll Free 1-800-822-8158
(24-hour telephone service available.)

or write to TAB Books, Blue Ridge Summit, PA 17294-0840.

Title	Product No.	Quantity	Price

☐ Check or money order made payable to TAB Books

Charge my ☐ VISA ☐ MasterCard ☐ American Express

Acct. No. _____ Exp. _____

Signature: _____

Name: _____

Address: _____

City: _____

State: _____ Zip: _____

	Subtotal	$ _____
	Postage and Handling ($3.00 in U.S., $5.00 outside U.S.)	$ _____
	Add applicable state and local sales tax	$ _____
	TOTAL	$ _____

TAB Books catalog free with purchase; otherwise send $1.00 in check or money order and receive $1.00 credit on your next purchase.

Orders outside U.S. must pay with international money in U.S. dollars drawn on a U.S. bank.

TAB Guarantee: If for any reason you are not satisfied with the book(s) you order, simply return it (them) within 15 days and receive a full refund.

BC